W0232530

Study Guide for Education to Accompany Salkind and Frey's

Statistics for
People Who (*Think They*)
Hate Statistics

7
EDITION

Sara Miller McCune founded SAGE Publishing in 1965 to support the dissemination of usable knowledge and educate a global community. SAGE publishes more than 1000 journals and over 800 new books each year, spanning a wide range of subject areas. Our growing selection of library products includes archives, data, case studies and video. SAGE remains majority owned by our founder and after her lifetime will become owned by a charitable trust that secures the company's continued independence.

Los Angeles | London | New Delhi | Singapore | Washington DC | Melbourne

Study Guide for Education to Accompany Salkind and Frey's

Statistics for People Who *(Think They)* Hate Statistics

7
EDITION

Neil J. Salkind
Bruce B. Frey
The University of Kansas

Prepared by Susan Parault Dowds
St. Cloud State University

Los Angeles | London | New Delhi
Singapore | Washington DC | Melbourne

FOR INFORMATION

SAGE Publications, Inc.
2455 Teller Road
Thousand Oaks, California 91320
E-mail: order@sagepub.com

SAGE Publications Ltd.
1 Oliver's Yard
55 City Road
London, EC1Y 1SP
United Kingdom

SAGE Publications India Pvt. Ltd.
B 1/I 1 Mohan Cooperative Industrial Area
Mathura Road, New Delhi 110 044
India

SAGE Publications Asia-Pacific Pte. Ltd.
18 Cross Street #10-10/11/12
China Square Central
Singapore 048423

Copyright © 2020 by SAGE Publications, Inc.

All rights reserved. No part of this book may be reproduced or utilized in any form or by any means, electronic or mechanical, including photocopying, recording, or by any information storage and retrieval system, without permission in writing from the publisher.

All trademarks depicted within this book, including trademarks appearing as part of a screenshot, figure, or other image are included solely for the purpose of illustration and are the property of their respective holders. The use of the trademarks in no way indicates any relationship with, or endorsement by, the holders of said trademarks. SPSS is a registered trademark of International Business Machines Corporation.

Printed in the United States of America

ISBN 978-1-5443-9597-5

Acquisitions Editor: Helen Salmon
Editorial Assistant: Megan O'Heffernan
Production Editor: Jane Martinez
Typesetter: Hurix Digital
Proofreader: Christine Dahlin
Cover Designer: Candice Harman
Marketing Manager: Shari Countryman

This book is printed on acid-free paper.

19 20 21 22 23 10 9 8 7 6 5 4 3 2 1

Contents

General Outline

Following is a general outline showing the sequence of content presented for each chapter.

Chapter Outline

Learning Objectives

Summary/Key Points

Key Terms

True/False Questions

Multiple-Choice Questions

Exercises

Short-Answer/Essay Questions

SPSS* Questions

Just for Fun/Challenge Yourself

Answer Key

---

* IBM® SPSS® Statistics is a registered trademark of International Business Machines Corporation.

A Note about the Dataset Accompanying this Study Guide

This dataset can be found on the website at **edge.sagepub.com/salkindfrey7e** by clicking on **Resources for the Study Guide for Education** on the left-hand navigation menu.

The Teacher Survey Data used in many of the SPSS exercises for the Study Guide for Education is simulated data. The variables in the data set are detailed below:

Variable	Label	Values	Level of Measurement
Teacher Type	Category of teacher	1 = Traditional (teaches in English) 2 = Spanish Immersion 3 = Special Education 4 = Art or Music 5 = Health or Physical 6 = English Language Teacher (teaching ELL students) 7 = Specialists (e.g. Speech or Reading) 8 = Gifted or High Potential	Nominal
Years of Experience	Years of teaching experience		Ratio
Grade Level	Current grade level taught	0 = Kindergarten, 1 = 1st, 2 = 2nd, 3 = 3rd, 4 = 4th, 5 = 5th 6 = multiple grade levels	Nominal (because value 6 is not equal to 6th grade)
Location	Number of teaching locations		Ratio
Degree	Highest degree obtained	1 = Bachelor's, 2 = Master's, 3 = PhD	Ordinal
License	Type of teaching license	1 = traditional, 2 = alternative, 3 = no license	Nominal
Professional Development	Hours of professional development over the last 12 months	1 = 8 or fewer hours 2 = 9–16 hours 3 = 17–24 hours 4 = more than 24 hours	Interval
Full Time	Full time status	1 = full time, 2 = part time	Nominal
Per Week Hours	Estimated hours worked per week		Ratio
Work Another Job	Do you work another job to supplement your salary?	1 = no 2 = yes, but only during the summer months 3 = yes, all year long	Nominal

The Teacher Survey Data is not real data collected from teachers. It is simulated data meant for educational purposes only. The data set contains simulated data to represent a random elementary school in the U.S. This fictitious elementary school consists of grades K–5 and has traditional classes taught in English as well as Spanish immersion classes. The simulated data represents a set of 70 teachers in this school.

The next items in the data set are from a survey of teacher satisfaction. The survey is on a five point scale with 1 = Strongly Disagree, 2 = Disagree, 3 = Neutral, 4 = Agree and 5 = Strongly Agree (this would be an interval scale). The following are the survey items:

1. The school administration's behavior towards the staff is supportive and encouraging.

2. My principal communicates with me often.

3. At this school teachers have a lot of influence over school policy.

4. Routine duties and paperwork interfere with my teaching.

5. There is a great deal of cooperation among staff members at this school.

6. I am given the support I need to teach students with special needs.

7. I have a lot of control over the teaching techniques I use in my class.

8. I receive a great deal of support from parents.

9. Poverty is a major issue at this school.

10. I am very stressed at work.

11. I am satisfied with my salary.

12. I am generally satisfied with being a teacher at this school.

The above 12 item survey was given at the midpoint of the school year (listed as SurveyItem A + the survey number in the data set). At the last week of the school year two of the survey questions were given again (listed as SurveyItemB in the data set). These items are

B10. I am very stressed at work.

B12. I am generally satisfied with being a teacher at this school.

The remaining three variables are combinations created by adding together similar survey items. The variable "School Administration" is a combination of survey items A1 and A2 which both refer to the administration's interactions with teachers. The variable "Overall Stress" is a combination of the "I am very stressed at work" item from midyear and the last week of school. The variable "Overall Satisfaction" is a combination of the "I am generally satisfied with being a teacher at this school" item from midyear and the last week of school.

Note: The teaching categories used in this data set by no means represent all teachers or staff in a school. Paraprofessional and other support staff have been left out only to simplify the data set.

1 Statistics or Sadistics? It's Up to You

LEARNING OBJECTIVES

- Understand the purpose and scope of statistics.
- Review (briefly) the history of statistics.
- Get an introduction to descriptive and inferential statistics.
- Review the benefits of taking a statistics course.
- Learn how to use and apply this book.

SUMMARY/KEY POINTS

Introduction to Part I

- Researchers in a very wide variety of fields use statistics to make sense of the large sets of data they collect in studying a great number of interesting problems.
  - Aletha Huston, a researcher and teacher, has found that children who watch educational programs on television do better in school than those who don't.
  - Sue Kemper, a professor of psychology, has studied the health of nuns, finding that the complexity of the nuns' writing during their early 20s is related to risk for Alzheimer's disease as many as 70 years later.
  - Michelle Lampl, a pediatrician and anthropologist, has studied the growth of infants, finding that some infants can grow as much as 1 inch overnight.

- Statistics can be defined as "the science of organizing and analyzing information," making that information easier to understand.

- Statistics are used to make sense of often large and unwieldy sets of data.

- Statistics can be used in education, or any field, to answer a very wide variety of research questions and hypotheses.

A Brief History of Statistics

- Far back in human history, collecting information became an important skill.

- Once numbers became part of human language, they began to be attached to outcomes. In the 17th century, the first set of data relating to populations of people was collected.

- Once sets of data began to be collected, scientists needed to develop specific tools to answer specific questions. This led to the development of statistics.

- In the early 20th century, the simplest test for examining the differences between the averages of two groups was developed and these have evolved into the more refined tests used every day in education settings

- The development of powerful and relatively inexpensive computers has revolutionized the field of statistics. While individuals can now conduct complex and computationally intensive statistical analyses with their own computers, they can potentially run analyses incorrectly or arrive at incorrect conclusions regarding their results.

- Today, researchers from a wide variety of fields use basically the same techniques, or statistical tests, to answer very different questions. This means that learning statistics

Chapter 1 ♦ Statistics or Sadistics? It's Up to You

3

enables you to conduct quantitative research in almost any field. This is important for educators who must understand multiple fields of study from psychology to technology and much more.

Statistics: What It Is (and Isn't)

- Statistics describes a set of tools and techniques that is used for describing, organizing, and interpreting information or data. It helps us understand the world around us.

- Descriptive statistics are used to organize and describe the characteristics of a collection of data. The collection is sometimes called a data set or just data.

- Inferential statistics are often (but not always) carried out after descriptive statistics. They are used to make inferences from a smaller group of data to a larger one. An example is using results from one kindergarten classroom to infer, or generalize, about a population of a whole kindergarten grade.

- A sample is a portion or subset of a larger population. Data from samples may be used for description only, or to generalize something about the larger population. Samples should always adequately represent the population in terms of age, ethnicity, gender, religion as well as any other relevant characteristic.

- A population is a full set from which a sample is taken: all the possible cases of interest. Data from a sample can be used to infer properties of a whole population.

Why Study Statistics? What Am I Doing in a Statistics Class?

- Having statistical skills puts you at an advantage when applying to graduate school or for a research or academic position.

- If not a required course for your major, a basic statistics course on your transcript sets you apart from other students.

- A statistics course can be an invigorating intellectual challenge.

- Having a knowledge of statistics makes you a better student, as it will enable you to better understand journal articles and books in education or other fields as well as what your professors and colleagues study and discuss.

- Research and statistical analyses are the foundation of "evidence based education." Understanding the statistics behind the evidence will allow you to understand why policy changes or curriculum changes are occurring.

- A basic knowledge of statistics will position you well for further study if you plan to pursue a graduate degree in education or other fields.

Tips for Using This Book

- Be confident: You are an educator, you can handle this. Work hard, and you'll do fine.

- Statistics is not as difficult as it's made out to be.

- Don't skip chapters: Work through them in sequence.

- Form a study group.

- Ask your professor questions.

- Do the exercises at the end of each chapter.

- Practice, practice, practice: Besides the exercises, find other opportunities to use what you've learned.

- Look for applications to make the material more real.

- Browse: Flip through the future material and review chapters.

- Have fun: Enjoy mastering a new field and acing your course.

KEY TERMS

- **Statistics**: A set of tools and techniques that is used for describing, organizing, and interpreting information or data

- **Descriptive statistics**: A set of statistical techniques and tools that is used to organize and describe data

- **Data, Data set**: A set of data points (where one data point = one observation/measurement)

- **Inferential statistics**: A set of statistical techniques and tools that is used to make inferences from a smaller group of data to a larger one

- **Sample**: A subset of the population. A researcher's goal is often to generalize findings from a sample to a population

- **Population**: All the possible subjects or cases of interest

2 Computing and Understanding Averages

Means to an End

CHAPTER OUTLINE

- ✦ Computing the Mean
- ✦ Computing the Median
- ✦ Computing the Mode
  - ✧ Apple Pie à la Bimodal
- ✦ When to Use What Measure of Central Tendency (and All You Need to Know About Scales of Measurement for Now)
  - ✧ A Rose by Any Other Name: The Nominal Level of Measurement
  - ✧ Any Order Is Fine With Me: The Ordinal Level of Measurement
  - ✧ 1 + 1 = 2: The Interval Level of Measurement
  - ✧ Can Anyone Have Nothing of Anything? The Ratio Level of Measurement
  - ✧ In Sum...
- ✦ Using SPSS to Compute Descriptive Statistics
  - ✧ The SPSS Output
- ✦ Real-World Stats
- ✦ Summary
- ✦ Time to Practice

LEARNING OBJECTIVES

- Understand averages, or measures of central tendency, one of the key components of descriptive statistics.

- Learn how to calculate the mean, median, and mode.

- Understand the distinction among these three measures of the average.

- Understand that the mean is very sensitive to outliers.

- Understand and apply scales or levels of measurement.

- Understand which measures of the average to use with different types of data.

- Learn how to use SPSS to calculate measures of central tendency.

SUMMARY/KEY POINTS

- Averages, or measures of central tendency, are used to determine the single value that best represents an entire group of scores. Popular measures of the average include the mean, median, and mode.

- The mean consists of the middle point of a set of values, and it is simply the sum of all the values divided by the number of values.

- The median consists of the middle point of a set of cases, and the mode represents the most frequent value in a set of scores.

- Only the mode can be used when determining an average for qualitative, categorical, or nominal data.

- Likewise, the median and mean can only be used with quantitative data.

- The mean can be considered the most precise measure, followed by the median and, finally, by the mode. While the mean is the most precise measure, it is sensitive to extreme scores. The median is not sensitive to extreme scores, so it better represents the center-most value of a data set that does include extreme scores.

- Scales of measurement are key to choosing the correct measure of central tendency to use. There are certain characteristics of data, specifically the scale of measurement, and each level of the scale builds upon the previous. The scale of measurement can be at the nominal, ordinal, interval, or ratio levels. The more precise levels of measurement (for example, the interval level of measurement) contain all the qualities of the scales below them.
  - **Nominal** variables are by class or category, are the least precise, and are mutually exclusive. Gender, ethnicity, political affiliation, students with an IEP vs. those without can all be nominal variables. **Ordinal** variables are measured in order or rank. The order is noted, but there is no way to tell the amount of difference between each rank. Class rankings, sports rankings, or finishers in a race could all be interval variables. **Interval** variables are frequently used in tests or assessment tools in which there is some underlying continuum that indicates the amount of a higher or lesser value. Intervals, or spaces or points, along a continuum are equal to one another. The number correct on a test could be used as an interval variable (e.g., if you get 10 correct on a 10-point vocabulary test, and someone else gets 5 correct, then you got twice as many correct). An assessment tool at the **ratio** level of measurement means that it has an absolute zero (e.g., weight or length, years of school). Ratio is the highest level of scales of measurement.

- The following are guides for when to use what scale of measurement (but there can be exceptions): Use the mode when variables are nominal. Use the median when you have extreme scores and you do not want to distort the average. Use the mean when you have data that do not include extreme scores and the variables are not categorical. The nominal level of measurement is the least precise, while the ratio level of measurement is the most precise. The "higher up" you are on the scale of measurement, the more precise, detailed, and informative your data are.

- For a sample statistic, Roman letters are used. For a population parameter, Greek letters are used.

KEY TERMS

- **Average**: The one value that best represents an entire group of scores. This can be the mean, median, or mode.

- **Measures of central tendency**: Another term for *averages*. As in the definition of average, measures of central tendency consist of the mean, median, and mode.

- **Mean**: The sum of all the values in a group, divided by the number of values in the group
  - The mean is sometimes represented using X-bar, or the letter *M*, and it is also called the typical, average, or most central score.
  - The mean is very sensitive to extreme scores.
  - When calculating the mean by hand, computing the "weighted mean" can save time when your data contain multiple instances of different values.

- **Median**: The midpoint of a set of scores
  - The median is sometimes abbreviated as *Med* or *Mdn*.
  - To compute the median, list all the values in order, from highest to lowest or lowest to highest. Next, find the middle-most score. If you have an even number of values, the median is calculated as the mean of the two middle values.
  - **Percentile points**: These are used to define the percentage of cases equal to and below a certain point in a distribution or set of scores. A score at the 25th percentile (Q_1) is at or above 25% of the other scores in the distribution. A score at the 50th percentile is often referred to as Q_2 and is the median. Percentile points are the basis for percentile rank scores on standardized education tests. In education the 50th percentile on a nationally standardized test is viewed as grade level. Scores below the 50th percentile are considered below grade level, those above the 50th percentile are considered above grade level, and scores at the 50th percentile are considered at grade level. For example, a student who obtained a percentile rank score of 40 on a 3rd grade standardized vocabulary measure scored as well as or better than 40% of students in the distribution of scores. This student's score is considered below grade level.

  - Because the median focuses on cases, and not the values of those cases, it is much less sensitive to extreme scores, or outliers, than is the mean.

- **Mode**: The value that occurs most frequently in a set of data
  - To find the mode, first list all the values in the distribution, listing each value only once. Next, count the number of times that each value occurs. The one that occurs most often is the mode.
  - If a set of values has more than one mode, the distribution is multimodal.
  - A distribution can be multimodal even if it has multiple modes that are very similar but not exactly the same (i.e., 15 of one category and 16 of some other category).

- **Skew (verb):** When your data include too many extreme scores, the distribution of scores can become *skewed*, or significantly distorted.

- **Data points:** Individual observations in a set of data

- **Scales of measurement:** Different levels at which outcomes are measured. The four scales of measurement are nominal, ordinal, interval, and ratio.
  - **Nominal level of measurement:** The level of measurement such that outcomes can only be placed into unranked categories
  - **Ordinal level of measurement:** The level of measurement such that outcomes can be rank ordered
  - **Interval level of measurement:** The level of measurement such that outcomes are based on some underlying continuum that makes it possible to speak about how much greater one performance is than another
  - **Ratio level of measurement:** The level of measurement such that outcomes are based on some underlying continuum that also contains a true, or absolute, zero

TRUE/FALSE QUESTIONS

1. The mode and median are both averages.

2. The mean is very sensitive to extreme scores.

3. The more precise levels of measurement (for example, the interval level of measurement) contain all the qualities of the scales below them.

4. Ordinal variables have two features. They show order or rank and they have equal distance between points along a scale.

5. Data on student ethnicity would be at the ratio scale of measurement.

MULTIPLE-CHOICE QUESTIONS

1. Of the set of values {170, 249, 523, 543, 572, 689, 1,050}, what is 543?

   a. The mean

   b. The median

   c. The mode

   d. The percentile

2. What is the mean of the following set of values: 1,501, 1,736, 1,930, 1,176, 446, 428, 768, and 861?

   a. 1,105.75

   b. 1,018.5

   c. 428

   d. 1,930

   e. 8

3. Your data set contains a variable on region of residence that contains the following possible responses: Northeast, South, Midwest, West, and Pacific Coast. Which of the following measures of central tendency should you use for this variable?

    a. The mean

    b. The median

    c. The mode

    d. The weighted mean

    e. Both a and b

4. This is the level of measurement where outcomes are based on some underlying continuum such that it is possible to speak about how much greater one performance is than another one.

    a. Nominal

    b. Ordinal

    c. Interval

    d. Ratio

5. This is the level of measurement such that outcomes can be placed only into unranked categories.

    a. Nominal

    b. Ordinal

    c. Interval

    d. Ratio

6. This is the level of measurement such that outcomes are based on an underlying continuum that contains a true, or absolute, zero.

    a. Nominal

    b. Ordinal

    c. Interval

    d. Ratio

7. This is the level of measurement such that outcomes can be rank ordered.

    a. Nominal

    b. Ordinal

    c. Interval

    d. Ratio

8. This is the most precise level of measurement.

    a. Nominal

    b. Ordinal

    c. Interval

    d. Ratio

9. This is the least precise level of measurement.

    a. Nominal

    b. Ordinal

    c. Interval

    d. Ratio

10. The number of Russian words a student knows is on what scale of measure?

    a. Nominal

    b. Ordinal

    c. Interval

    d. Ratio

11. A 5th grade teacher gave a test on scientific classifications. She has two students in her class that were recently tested for the gifted program and meet the criteria for this program. These two students are likely to

    a. increase the median of the total test scores.

    b. increase the mean of the total test scores.

    c. increase the mode of the total test score.

    d. decrease the mode of the total test score.

EXERCISES

1. Calculate the mean for the following set of values: 25, 37, 53, 72, 76.

2. Calculate the median for the same set of values: 25, 37, 53, 72, 76.

3. Calculate the mode for the following set of values: 7, 7, 12, 15, 17, 19, 22, 25, 27, 31, 35, 42, 42, 47, 59.

4. An AP psychology teacher gets the results of this year's AP test. The scores are as follows: 1, 3, 4, 2, 5, 2, 3, 2, 3, 4, 3, 3, 2, 4, 2, 1, 5. Calculate the mean, median and mode for this set of scores.

SHORT-ANSWER/ESSAY QUESTIONS

1. Review the following set of values: 12, 24, 37, 42, 55, 62, 72, 77, 246, 592. What would be the best measure of central tendency to use for this set of values? Why?

2. You have just finished conducting a study on students that contained a large set of questions on demographic information, including about the participants' gender, year in school, and race. What measure of central tendency should be used for these types of variables? Why?

3. In a survey that asked students about their favorite styles of music, 37 replied rock, 27 said they preferred pop music, 14 stated they liked rap music the most, and 3 individuals

stated that they preferred country. Based on these data, what would be the mode? Why couldn't the mean or median be used to describe this set of values?

4. You just took the GRE (in preparation for graduate school) and received a total score in the 98th percentile. Is this a good or bad score? Explain.

5. Come up with examples of nominal, ordinal, interval, and ratio levels of measurement in education.

SPSS QUESTIONS

1. Input the following set of 25 scores into SPSS: 72, 13, 79, 76, 29, 8, 12, 27, 90, 72, 29, 40, 22, 45, 28, 50, 40, 84, 71, 14, 56, 46, 25, 28, 33. Use SPSS to calculate the mean, median, and mode.

   Now, go through the numbers yourself and calculate these averages by hand. Make sure your calculations match those produced by SPSS.

2. Open the supplemental data set "Teacher Survey Data" in SPSS. Use SPSS to calculate the mean, median, and mode for the variable "years of teaching experience."

JUST FOR FUN/CHALLENGE YOURSELF

1. A set of five values was found to have a mean of 54.8. You know that four of the values are 24, 27, 53, and 68. What is the fifth value?

2. It is close to the end of the semester, and your goal is to get at least an A– (an average of 90) in your class. Each of your three exams is worth 25% of your grade, and your final is also worth 25%. You received an 87 on your first exam, an 88 on your second exam, and a 91 on your third exam. What is the minimum score (represented as a whole number) that you need to get on your final in order to get your A–?

ANSWER KEY

TRUE/FALSE QUESTIONS

1. True. The mean, median, and mode are all different averages.

2. True.

3. True.

4. False. Ordinal variables only show order or rank. Interval variables have equal distance between points along a scale.

5. False. The variable of ethnicity would be a nominal variable.

MULTIPLE-CHOICE QUESTIONS

1. (b) The median
2. (a) 1,105.75
3. (c) The mode
4. (c) Interval
5. (a) Nominal
6. (d) Ratio
7. (b) Ordinal
8. (d) Ratio
9. (a) Nominal
10. (d) Ratio
11. (b) increase the mean.

EXERCISES

1. 52.6
2. 53
3. Modes = 7 and 42.
4. Mean = 2.88, median = 3, mode = 2 and 3 (multiple modes)

SHORT-ANSWER/ESSAY QUESTIONS

1. Based on this set of values, the best measure of central tendency would be the median. First, the mode is only preferred in situations in which your variable is qualitative or categorical (i.e., situations in which the mean or median cannot be computed). Second, the mean should not be used in this situation because you have two very high outliers, 246 and 592.

2. For these types of variables, you can only use the mode. Since these variables are categorical/ qualitative, it would be impossible to calculate the mean. Likewise, the median cannot be computed for variables of this nature.

3. Based on these data, the mode would be "rock music," as this is the most common category of response. The mean or median could not be used with this type of variable as it is a qualitative/categorical variable.

4. This score is very good: Only 2% of people who took this exam did better than you did.

5. A nominal level of measurement is a categorical level of measurement in which outcomes fit into one and only one class. Some examples in education would be (1) students on free and reduced lunch and those who are not, or (2) parents' marital status: married, divorced, single. An ordinal level of measurement is any variable that is categorical (consisting of a number of discrete categories) and that can be ordered, or ranked. Some examples include highest degree earned, social class, and letter grade. An interval level of measurement is any variable that is continuous but does not have an absolute or true zero. Some examples are IQ, height, weight, and grade on an exam (in these examples, it is assumed there is no "true" zero). A ratio level of measurement involves the presences of absolute 0. Some examples in education would be the number of times a student has received detention and the amount of money in a student's lunch account.

SPSS QUESTIONS

1. Statistics

Var	
Valid	25
Missing	0
Mean	43.5600
Median	40.0000
Mode	28.00a

Multiple modes exist.

The smallest value is shown.

Mean = 43.56, Median = 40, Modes = 28, 29, 40, 72

2. Years of teacher experience

Var	
Valid	70
Missing	0
Mean	10.5286
Median	9.0000
Mode	4.00

JUST FOR FUN/CHALLENGE YOURSELF

1. $\bar{X} = \dfrac{\sum X}{n} \Rightarrow \dfrac{24+27+53+68+x}{5} = 54.8 \Rightarrow \dfrac{172+x}{5} = 54.8 \Rightarrow 172+x = 274 \Rightarrow x = 102.$

2. $\bar{X} = \dfrac{\sum X}{n} \Rightarrow \dfrac{87+88+91+x}{4} = 90 \Rightarrow \dfrac{266+x}{4} = 90 \Rightarrow 266+x = 360 \Rightarrow x = 94.$

3 Understanding Variability

Vive la Différence

LEARNING OBJECTIVES

- Understand what measures of variability are, how they are used, and how they differ from one another.

- Learn how to calculate the range, standard deviation, and variance by hand.

- Learn how to use SPSS to calculate measures of variability.

- Understand the benefits of unbiased estimates.

SUMMARY/KEY POINTS

- In addition to measures of central tendency, measures of variability make up another important component of descriptive statistics.
  - Variability is a measure of how much each score in a group of scores differs from the mean.

- Measures of variability include the range, standard deviation, variance, and mean deviation.
  - The range is the easiest measure of variability to calculate, but it is the most general. It should never be used alone to reach any conclusions regarding variability.
  - The standard deviation, the most frequently used measure of variability, is a measure of the average distance from the mean.
  - The variance is the standard deviation squared. It is seldom reported in journal articles and is primarily used in statistical formulas as a practical measure of variability.

- Both the standard deviation and the variance include the term $n - 1$. Estimates in which 1 is subtracted from the sample size are called *unbiased estimates* and are more conservative estimates of population parameters.

- One example of variability in education is the student's knowledge level. In general, it is easier to teach a lesson to a class with little variability in knowledge levels. When there is a lot of variability in knowledge levels the teacher must decide how to structure a lesson for those with very little knowledge of the content being taught while simultaneously not boring those with a lot of knowledge of the content. The teacher must adjust the lesson plan for the varying knowledge levels of the learners. This is why it is important to know both the mean level of knowledge as well as the variability in knowledge levels.

KEY TERMS

- **Variability**: The amount of spread or dispersion in a set of scores

- **Range**: The highest minus the lowest score. This is a very general estimate of the range.
  - **Exclusive range**: The highest score minus the lowest score
  - **Inclusive range**: The highest score minus the lowest score plus 1

- **Standard deviation**: The average amount of variability in a set of scores, or a measure of the average distance from the mean. This is the most common measure of variability, but it is sensitive to extreme scores.

- **Variance**: The square of the standard deviation. This measure is much more difficult to interpret than the standard deviation, which is why it is used much less often.

- **Mean deviation**: The sum of the absolute value of the deviations from the mean divided by the number of scores. This calculation differs from that of the standard deviation.

- **Unbiased estimate**: An estimate of a population parameter in which 1 is subtracted from n. It is considered to be a more conservative estimate than the biased estimate.
  - Biased estimates can be used if you only wish to describe a sample, while unbiased estimates should be used when making estimates of population parameters.

TRUE/FALSE QUESTIONS

1. Two sets of values that have the same mean must also have the same variability.

2. It is possible for two or more sets of values to have the same standard deviation and variance.

3. Before the standard deviation or variance is calculated, the mean of scores needs to be calculated.

4. Like the mean, the standard deviation is sensitive to extreme scores.

5. It is common for a set of scores to have no variability.

6. The variance is easier to interpret than the standard deviation.

7. In education, determining the mean is more important than determining the standard deviation.

MULTIPLE-CHOICE QUESTIONS

1. Which of the following sets of values has the greatest variability?

   a. 2, 2, 3, 3, 4

   b. 7, 7, 8, 9, 9

   c. 2, 3, 5, 7, 8

   d. 1, 4, 7, 9, 11

2. In which of the following sets of values is the mean equal to the standard deviation?

   a. 2, 4, 6, 8, 9

   b. 4, 7, 8, 8, 12

   c. 0, 2, 4, 6, 13

   d. 2, 4, 5, 9, 11

3. Which of the following sets of values has no variability at all?

   a. 10, 20, 30, 40, 50

   b. 2, 4, 6, 8, 10

   c. 0, 0, 0, 0, 5

   d. 8, 8, 8, 8, 8

4. What is the most general measure of variability?

   a. The range

   b. The standard deviation

   c. The variance

   d. The mean deviation

5. You are conducting a survey, and you wish to use a sample of respondents in order to estimate population parameters. Which of the following should you use?

   a. Biased estimates

   b. Unbiased estimates

   c. Both unbiased and biased estimates

6. Which of the following statements is correct?

   a. The range is typically equal to the mean.

   b. The range is equal to the square of the variance.

   c. The variance is equal to the square of the standard deviation.

   d. The standard deviation is equal to the square of the variance.

7. You are going on a Spring Break trip with your 8th grade class. Which of the following weather forecasts will require you to bring a sweater?

   a. A high temperature of 95 degrees with a range of 35 degrees

   b. A high temperature of 85 degrees with a range of 15 degrees

   c. A high temperature of 90 degrees with a range of 15 degrees

   d. A high temperature of 80 degrees with a range of 10 degrees

EXERCISES

1. Compute the range of the following set of scores: 214, 246, 379, 420.

2. Compute the standard deviation and variance of the following scores: 32, 48, 55, 62, 71.

SHORT-ANSWER/ESSAY QUESTIONS

1. When and why are unbiased estimates preferred over biased estimates?

2. Every year your school district sends out satisfaction surveys to parents. Your principal is planning to present this year's survey results at a staff meeting but will only give the staff the mean scores for the survey questions. Explain to your principal why it is also important to give the standard deviation scores for the data.

SPSS QUESTION

1. Input the following set of scores into SPSS: 68, 9, 17, 87, 19, 37, 66, 78, 33, 29, 33, 61, 3, 57, 50. Calculate the range, standard deviation, and variance of these scores using SPSS.

2. Open the supplemental data set "Teacher Survey Data" in SPSS. Use SPSS to calculate the range, standard deviation, and variance, for the variable "Per Week Hours," which is the teachers' estimate of how many hours of work they put in per week.

JUST FOR FUN/CHALLENGE YOURSELF

1. Calculate the mean deviation of the following set of scores: 23, 45, 57, 62, 88.

2. What are all the possible scenarios in which the variance of a set of values would equal the standard deviation?

ANSWER KEY

TRUE/FALSE QUESTIONS

1. False. In fact, it is much more likely that these two sets of scores have different measures of variability.

2. True. This is possible, but it would be very rare. This would be the case only if the standard deviation and variance were both equal to 0 or 1.

3. True. The equation for both the standard deviation as well as the variance includes the mean, so this value must be calculated before the standard deviation or variance is calculated.

4. True. Both the mean and the standard deviation are sensitive to extreme scores, also known as outliers.

5. False. It is extremely rare for a set of scores to have zero variability.

6. False. This is because the standard deviation is stated in the original units from which it was calculated, while the variance is stated in squared units.

7. False. Both the mean and the standard deviation are important scores in education (or any field) as they tell you two different things about a set of scores. The mean tells you the average of the scores and the standard deviation tells you the variability in the scores.

MULTIPLE-CHOICE QUESTIONS

1. (d) 1, 4, 7, 9, 11

2. (c) 0, 2, 4, 6, 13

3. (d) 8, 8, 8, 8, 8

4. (a) The range

5. (b) Unbiased estimates

6. (c) The variance is equal to the square of the standard deviation.

7. (a) A high temperature of 95 degrees with a range of 35 degrees.

EXERCISES

1. The range = 420 − 214 = 206.

2. First, $\bar{X} = \dfrac{\sum X}{n} \Rightarrow \dfrac{32+48+55+62+71}{5} = 53.6$.

   Next, the standard deviation =

   $$s = \sqrt{\frac{\sum(X-\bar{X})^2}{n-1}}$$

   $$= \sqrt{\frac{(32-53.6)^2+(48-53.6)^2+(55-53.6)^2+(62+53.6)^2+(71-53.6)^2}{5-1}}$$

   $$= 14.8.$$

   Finally, the variance = $s^2 = 14.8^2 = 219.04$.

SHORT-ANSWER/ESSAY QUESTION

1. Unbiased estimates are preferred in situations in which a sample is used to estimate population parameters. In these situations, unbiased estimates give you a more conservative estimate than do biased estimates. For example, when 1 is subtracted from the sample size, the standard deviation is forced to be larger than it would be otherwise, giving a more conservative estimate.

2. You should explain to your principal that the mean only tells the staff the average score and leaves out information on how much parents vary in their rates of satisfaction. For instance the survey could show a mean satisfaction rate of 4.8 (out of 6) on parent satisfaction with the school bullying policy with a standard deviation of 0.8. The survey may show the same mean of 4.8 (out of 6) on parent satisfaction with the school's use of technology but the standard deviation could be 2.5. Thus, overall parents have fairly high rates of satisfaction with both the bullying policy and the use of technology but parents vary more on their satisfaction with the use of technology. Thus, when it comes to bullying almost all parents show the same level of satisficed, however, when it comes to technology use some parents have very high rates of satisfaction and some much lower rates of satisfaction. This suggests then that the school needs to be either more consistent in their use of technology or further investigate why some parents are so satisfied while others are much less satisfied.

SPSS QUESTION

Descriptive Statistics				
	N	Range	Std. Deviation	Variance
Var	15	84.00	27.51069	756.838
Valid *N* (listwise)	15			

Descriptive Statistics				
	N	**Range**	**Std. Deviation**	**Variance**
Var	70	28	6.11714	37.419
Valid *N* (listwise)	70			

JUST FOR FUN/CHALLENGE YOURSELF

1. Calculate the mean deviation of the following set of scores: 23, 45, 57, 62, 88.

   First,

   $$\bar{X} = \frac{\sum X}{n} \Rightarrow \frac{23+45+57+62+88}{5} = 55.$$

   Next, the mean deviation

   $$= \frac{\sum|X - \bar{X}|}{n}$$

   $$= \frac{|23-55| + |45+55| + |57+55| + |62+55| + |88-55|}{5}$$

   $$= 16.8.$$

2. For this to be the case, you would need the square of the standard deviation (i.e., the variance) to be equal to the standard deviation. In other words, you would need a number that does not change when it is squared. This leaves you with only two values: 0 and 1. The only two possible scenarios in which the variance would be equal to the standard deviation is when both of these values are 0 or when both of these values are 1.

4 Creating Graphs

A Picture Really Is Worth a Thousand Words

LEARNING OBJECTIVES

- Learn how to create visually appealing and useful representations of data.

- Review different ways in which data can be represented using tables and graphs.

- Be able to choose the best type of chart based on the nature of your data.

- Learn how to draw several graphs by hand, as well as how to create graphs using SPSS.

SUMMARY/KEY POINTS

- The two previous chapters focused on measures of central tendency and variability. This chapter expands upon this, illustrating how differences in these measures result in different-looking distributions.

- A visual representation of data can be much more effective than numerical values alone at illustrating the characteristics of a distribution or data set.
  - Tables can be used to illustrate the distribution of a variable.
  - Tables covered here include frequency distributions and cumulative frequency distributions.
  - Charts/graphs can be used to pictorially represent the distribution of a variable.
  - Charts/graphs covered here include histograms, frequency polygons, bar charts, column charts, line charts, and pie charts.

- It's easy to build a "bad" chart. Certain guidelines should be followed to make a chart that is easy to read and clearly illustrates what you are trying to show.
  - Less is more—minimize "chart junk."
  - Label everything.
  - Communicate only one idea.
  - Maintain the scale in a graph (in a 3:4 ratio).
  - A chart alone should convey what you want to say.
  - Simple is best; limit the number of words.

KEY TERMS

- **Frequency distribution**: A method of tallying and representing how often certain scores occur. Frequency distributions generally group scores into class intervals or ranges of numbers.

- **Class interval**: A range of numbers, chosen by the researcher, to be used in charts/graphs

- **Histogram**: A graphical representation of a frequency distribution in which the frequencies are represented by bars

- **Midpoint**: The central point of a class interval

- **Frequency polygon**: A continuous line that represents a frequency distribution

- **Cumulative frequency distribution**: A frequency distribution that shows frequencies for class intervals along with the cumulative frequency for each
  - **Ogive**: Another name for a cumulative frequency polygon

- **Column charts**: A type of chart in which categories are organized horizontally on the x-axis and values are shown vertically on the y-axis. This type of chart is used to compare the frequencies of different categories with one another.

- **Bar charts**: A type of chart in which categories are organized vertically on the *y*-axis and values are shown horizontally on the *x*-axis in the form of separate, separated bars

- **Line chart**: A type of chart in which categories are organized vertically on the *y*-axis and values are shown horizontally on the *x*-axis. These values are connected by one or more lines.

- **Pie chart**: A type of chart that illustrates the proportions of responses to an item as a series of wedges in a circle

TRUE/FALSE QUESTIONS

1. A bar chart tallies and represents how often certain scores occur in the form of a table.

2. Skewness is a measure of the central point of a class interval.

3. A frequency polygon can be defined as a continuous line that represents a frequency distribution.

4. When creating a graph, everything should be labeled, and the graph should only communicate one idea.

5. Graphs should contain as much text as possible.

MULTIPLE-CHOICE QUESTIONS

1. When creating a graph, the ratio of the width to the length should be approximately _____.

   a. 1:2

   b. 2:1

   c. 3:4

   d. 5:7

2. When determining class intervals, you should aim to have about this many intervals cover the entire range of your data.

   a. 1 or 2

   b. 5 to 10

   c. 10 to 20

   d. 50 to 100

3. The largest class interval is placed _____ in a frequency distribution.

   a. at the top

   b. in the middle

   c. at the bottom

   d. randomly

4. When you want to compare the frequencies of different categories with one another, you should use a _____.

   a. pie chart

   b. line chart

   c. column chart

   d. frequency distribution

5. When you want to show a trend in the data at equal intervals, you should use a _____.

   a. pie chart

   b. line chart

   c. column chart

   d. frequency distribution

6. When you want to show the proportion of an item that makes up a series of data points, you should use a _____.

   a. pie chart

   b. line chart

   c. column chart

   d. frequency distribution

EXERCISES

1. Draw a histogram using the data from the following frequency distribution:

Class Interval	Frequency
90–100	2
80–89	8
70–79	4
60–69	12
50–59	14
40–49	20
30–39	14
20–29	7
10–19	3
0–9	1

2. Make the histogram you just created into a frequency polygon.

3. Using the frequency distribution from question 1 in this section, add an additional column to make it into a cumulative frequency distribution.

SHORT-ANSWER/ESSAY QUESTIONS

1. Measures of central tendency and variability describe a group of data and how different scores are from each other. However, visual representations of data offer a more effective way to examine the characteristics of a distribution or data set. What are some different ways in which charts, graphs, and figures can illustrate data?

2. Different types of charts or graphs work better for different types of variables. What are three guidelines to keep in mind, or questions to ask yourself, when choosing which chart or graph to use?

3. You are studying the change over time in the number of course credits needed to graduate with a Bachelor of Arts degree in Education. You have collected data on 3 universities at 10-year intervals over a 40-year span. What is the best chart or graph to illustrate your findings? Why?

SPSS QUESTIONS

1. Create a histogram using the data from question 1 in the "Exercises" section.

2. Create a pie chart using the following data on the student population of a school: Traditional English students = 422, Spanish immersion students = 389, Special Education = 37, Gifted = 41, ELL = 89.

3. Open the supplemental data set "Teacher Survey Data" in SPSS. Use SPSS to create a bar chart for the variable "professional development."

JUST FOR FUN/CHALLENGE YOURSELF

1. Using the cumulative frequency distribution you came up with for question 3 under the "Exercises" section, chart the cumulative frequency data as a cumulative frequency polygon, or ogive.

ANSWER KEY

TRUE/FALSE QUESTIONS

1. False. This describes a frequency distribution.

2. False. This describes a midpoint.

3. True.

4. True. This is good advice to follow when creating a graph.

5. False. Including too many words can detract from the visual message your chart is intended to convey to readers.

MULTIPLE-CHOICE QUESTIONS

1. (c) 3:4
2. (c) 10 to 20
3. (a) at the top
4. (c) column chart
5. (b) line chart
6. (a) pie chart

EXERCISES

1. Your histogram should look like this:

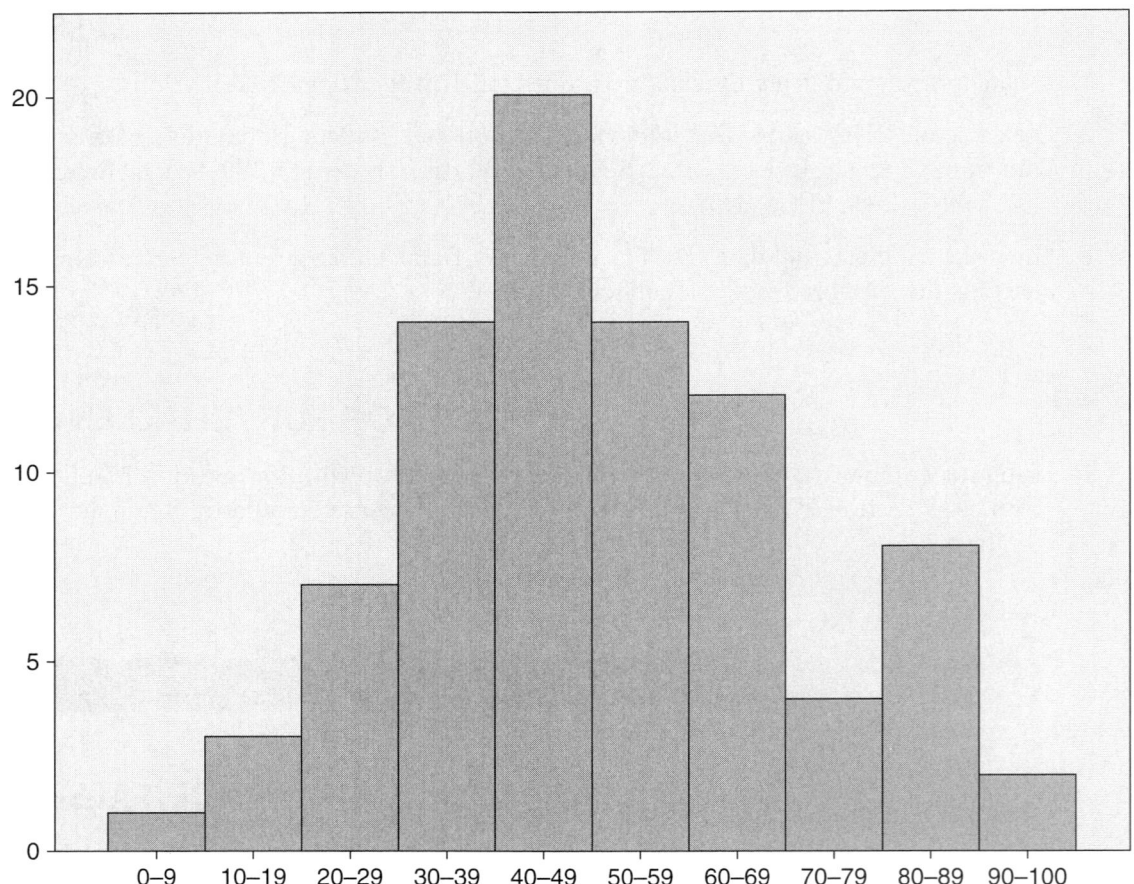

2. The frequency polygon should look like this:

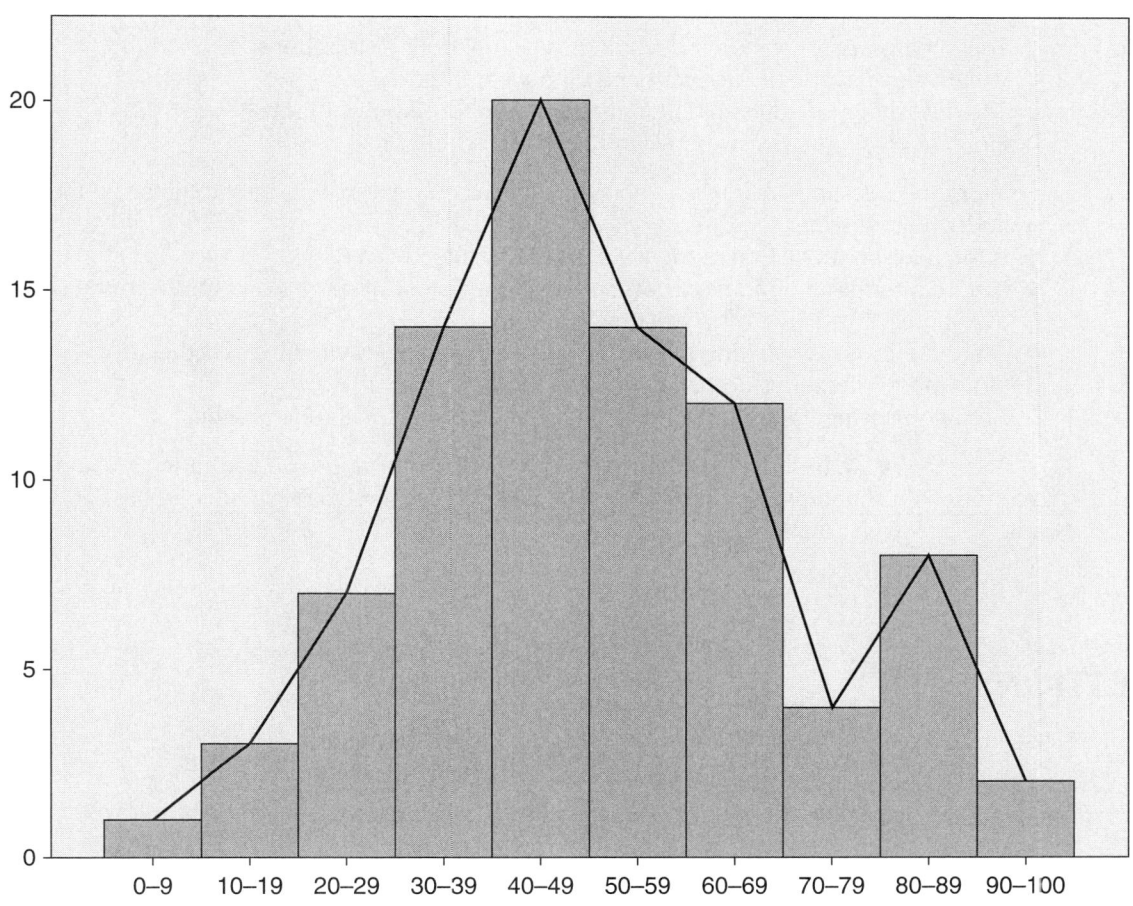

3. Your cumulative frequency distribution should look like this:

Class Interval	Frequency	Cumulative Frequency
90–100	2	85
80–89	8	83
70–79	4	75
60–69	12	71
50–59	14	59
40–49	20	45
30–39	14	25
20–29	7	11
10–19	3	4
0–9	1	1

SHORT-ANSWER/ESSAY QUESTIONS

1. Charts and graphs can illustrate the following:
   - That means and/or standard deviations have different distributions
   - What values of a variable occur and with what frequency
   - The data in a more dynamic manner than numbers alone can show
   - How much overlap exists between multiple distributions

2. Some of the questions that researchers may ask when determining how to illustrate their data include the following:
   - Is the measure of central tendency a mean, median, or mode?
   - Am I capturing one moment in time or telling a story about a trend for the same subjects over time?
   - Am I interested in showing the proportions of one category relative to others?
   - How many categories do I want to show at once?
   - Am I comparing the frequencies of different categories with one another?

3. A line chart is the best figure to illustrate your findings, because you can indicate the different schools with different lines and show changes in the number of credits (y-axis) at 10-year intervals (x-axis).

SPSS QUESTIONS

1. See the "Exercises" section, question 1, for the correct histogram.

2. The pie chart should look like this:

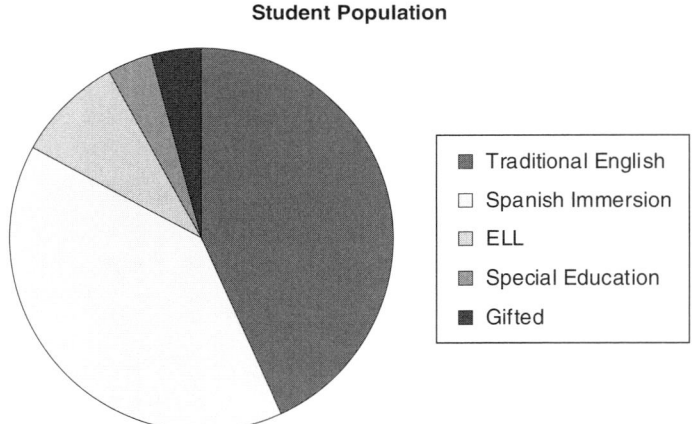

Student Population

■ Traditional English
☐ Spanish Immersion
▨ ELL
▨ Special Education
■ Gifted

3. The bar chart should look like this:

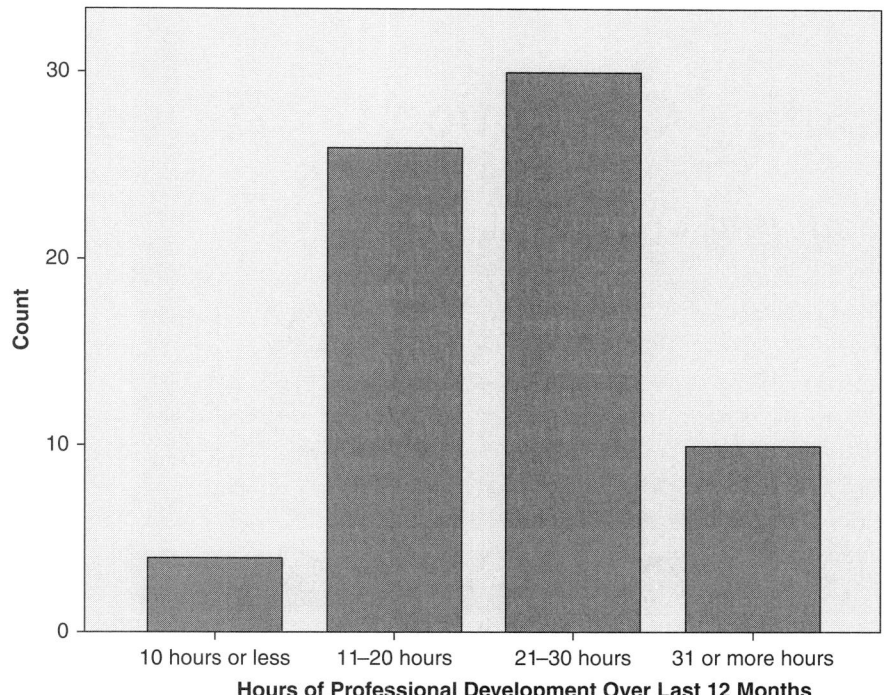

JUST FOR FUN/CHALLENGE YOURSELF

1. Your cumulative frequency polygon, or ogive, should look like the following image. Drawing the bars can be helpful if you are creating the cumulative frequency polygon by hand, but the bars are not necessary in an ogive.

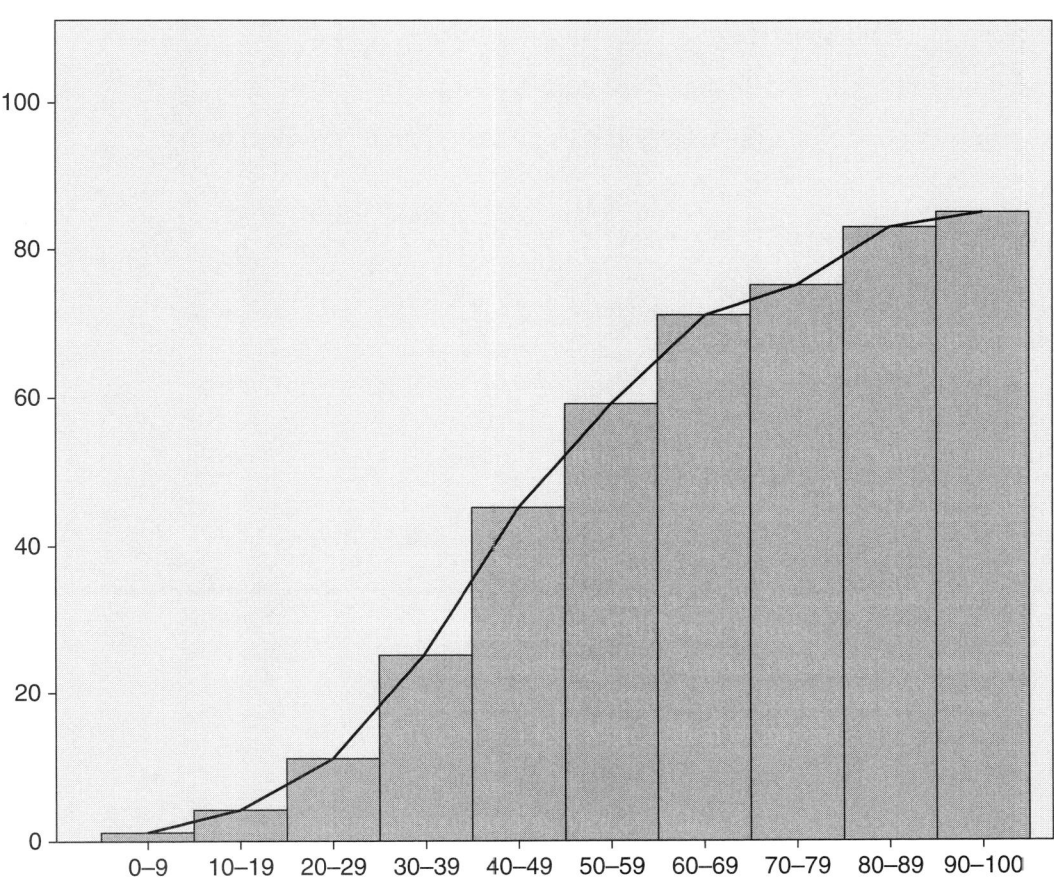

5 Computing Correlation Coefficients

Ice Cream and Crime

LEARNING OBJECTIVES

- Understand what correlation coefficients are used for.

- Learn how to interpret correlation coefficients.

- Be able to calculate Pearson's correlation coefficient by hand, as well as with SPSS.

- Be able to select the appropriate correlation coefficient to use depending on the nature of your variables.

SUMMARY/KEY POINTS

- Correlation coefficients are used to measure the strength and nature of the relationship between two variables.

- Pearson's correlation coefficient (r), the focus of this chapter, is used to calculate the correlation between two continuous (interval) variables.

- Other correlation coefficients can be used when one or more variables are ordinal or nominal.

- When you have a direct correlation—also known as a positive correlation—both variables change in the same direction. With an indirect correlation—also known as a negative correlation—variables change in opposite directions.

- Correlation coefficients focus on generalities. This means that the correlation that you find describes the group, not every individual person in your data.

- The absolute value of the correlation coefficient reflects the strength of the correlation. Coefficients can range from –1 to +1, and the closer a coefficient's absolute value is to 1, the stronger the relationship.

- The correlation between two variables will be reduced if the range of one or both of the variables is restricted.

- A scatterplot, or scattergram, can be used to visually illustrate a correlation between two variables. A positive slope represents a direct correlation, while a negative slope represents an indirect correlation.

- Correlation matrices are used to summarize correlations between a set of variables.

- The absolute value of a correlation corresponds to its strength—and corresponding meaningfulness—roughly as follows:
  - .8 to 1.0 is a very strong relationship.
  - .6 to .8 is a strong relationship.
  - .4 to .6 is a moderate relationship.
  - .2 to .4 is a weak relationship.
  - .0 to .2 is a weak or no relationship.

- The coefficient of determination is more precise than a correlation coefficient alone. The coefficient of determination is equal to the percentage of variance in one variable that is accounted for by the variance in a second variable.

- The fact that two variables are correlated does not imply that one causes the other.

KEY TERMS

- **Correlation coefficient**: A numerical index that reflects the relationship between two variables
  - Its range is from –1 to +1.
  - It is also known as a bivariate correlation.

- **Pearson product-moment correlation**: A specific type of correlation coefficient developed by Karl Pearson. It is specifically suited to determining the correlation between two continuous variables.

- **Direct correlation**: A positive correlation such that the values of both variables change in the same direction

- **Indirect correlation**: A negative correlation such that the values of the two variables move in opposite directions

- **Scatterplot** or **scattergram**: A plot of matched data points. This type of chart is used to illustrate a correlation between two variables.

- **Linear correlation**: A correlation that is best expressed as a straight line

- **Curvilinear relationship**: A situation in which the correlation between two variables begins as a direct correlation and then becomes an indirect correlation, or vice versa. It can be detected by examining the scatterplot.

- **Correlation matrix**: A table of correlation coefficients in which variables comprise the rows and columns of the table and the intersections of the variables are represented by correlation coefficients

- **Coefficient of determination**: The amount of variance accounted for in a relationship between two variables
  - It is equal to the square of the Pearson product-moment correlation coefficient.

- **Coefficient of alienation** (aka coefficient of nondetermination): The amount of unexplained variance in a relationship between two variables
  - It is equal to 1 minus the coefficient of determination.

- **Phi coefficient**: A measure used to estimate the correlation between two nominal variables

- **Rank biserial coefficient**: A measure used to estimate the correlation between one nominal and one ordinal variable

- **Point biserial coefficient**: A measure used to estimate the correlation between one nominal and one interval variable

- **Spearman rank coefficient**: A measure used to estimate the correlation between two ordinal variables

TRUE/FALSE QUESTIONS

1. If two variables are correlated, then one of the variables causes the other.

2. If your variables are found to be correlated, then the variables are correlated for the entire group of respondents but most likely not correlated for each individual case in your data set.

3. Arriving at a negative correlation is always a worse result than finding a positive correlation.

4. The correlation between time spent paying attention in class and an exam grade should be negative.

5. A correlation of .33 between amount of time spent on paperwork and teacher stress levels shows that the more paperwork a teacher has to do, the higher the level of stress.

MULTIPLE-CHOICE QUESTIONS

1. Which of the following is the range for Pearson's correlation coefficient?

   a. −10 to +10

   b. −5 to +5

   c. −1 to +1

   d. −100 to +100

2. Pearson's product-moment correlation can be used to calculate the correlation between these two types of variables.

   a. Two interval variables

   b. Two ordinal variables

   c. One nominal and one ordinal variable

   d. Two nominal variables

   e. All of the above

3. If one variable increases while the other decreases, you have this type of correlation.

   a. Direct correlation

   b. Indirect correlation

   c. Curvilinear correlation

   d. Leptokurtic correlation

4. If two variables move in the same direction (i.e., one increases as the other increases, and one decreases as the other decreases), you have a _____ correlation.

   a. direct

   b. indirect

   c. curvilinear

   d. leptokurtic

5. Which of the following is the strongest correlation?

   a. −.15

   b. +.27

    c. −.70

    d. +.55

6. Which of the following is the weakest correlation?

    a. −.22

    b. −.78

    c. +.12

    d. +.89

7. Pearson's product-moment correlation coefficient is represented by which of the following letters?

    a. r

    b. p

    c. t

    d. c

    e. z

8. If you compute the correlation between two variables, and one of the variables never changes, you can be sure that the Pearson correlation coefficient is equal to _____.

    a. +1

    b. 0

    c. −1

    d. +.5

    e. −.5

9. If a correlation is computed between two variables, but the range of one of the variables is restricted, your correlation will be _____.

    a. lower

    b. higher

    c. the same

    d. 0

    e. +1

10. In a scatterplot, if the dots cluster from the lower left-hand corner to the upper right-hand corner, the two variables have _____ correlation.

    a. a direct

    b. an indirect

    c. a curvilinear

    d. a bilinear

11. Which of the following would be considered a very strong correlation coefficient?

    a. .8 to 1.0

    b. .6 to .8

    c. .4 to .6

    d. .2 to .4

    e. .0 to .2

12. Which of the following would be considered a moderate correlation coefficient?

    a. .8 to 1.0

    b. .6 to .8

    c. .4 to .6

    d. .2 to .4

    e. .0 to .2

13. View the following scatterplot. What is the best estimate of the correlation between these two variables?

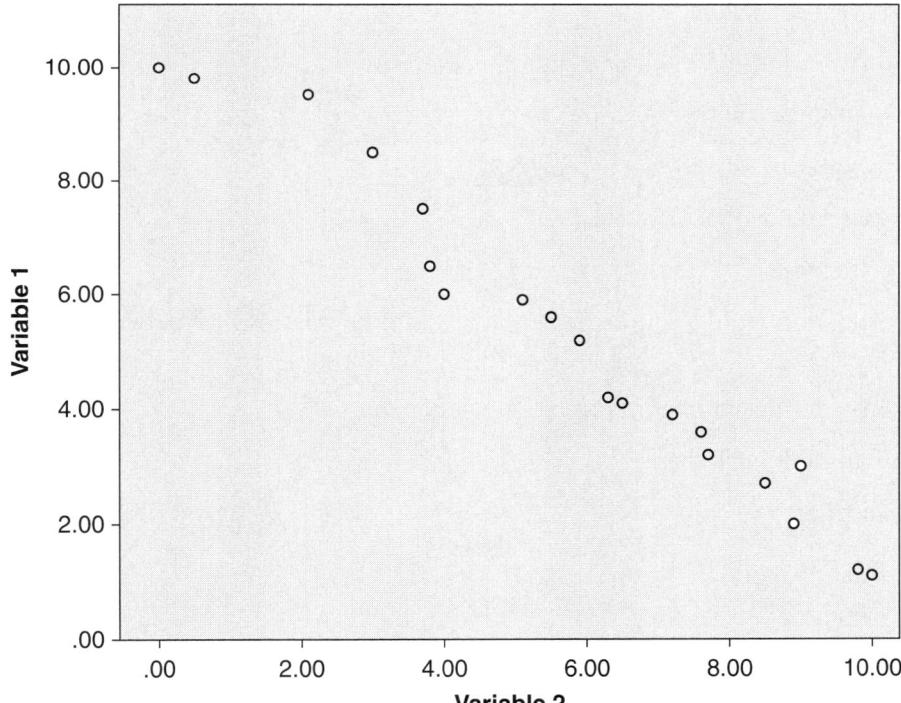

    a. −.20

    b. +.55

    c. −.98

    d. +.90

14. Which of the following correlation coefficients would indicate a weak or no relationship between two variables?

    a. .8 to 1.0

    b. .6 to .8

    c. .4 to .6

    d. .2 to .4

    e. .0 to .2

15. Which correlation would you use when analyzing the relationship between two nominal variables?

    a. Pearson's product-moment correlation coefficient

    b. Spearman rank coefficient

    c. Point biserial

    d. Rank biserial coefficient

    e. Phi coefficient

16. Which correlation would you use when analyzing the relationship between a nominal and an interval variable?

    a. Pearson's product-moment correlation coefficient

    b. Spearman rank coefficient

    c. Point biserial

    d. Rank biserial coefficient

    e. Phi coefficient

17. Which correlation would you use when analyzing the relationship between two ordinal variables?

    a. Pearson's product-moment correlation coefficient

    b. Spearman rank coefficient

    c. Point biserial

    d. Rank biserial coefficient

    e. Phi coefficient

18. A researcher found a correlation of .45 between parental involvement in education and elementary students' overall achievement levels. This suggests that

    a. parents are highly involved in their elementary school children's education.

    b. parental involvement has no effect on a child's educational achievement levels in elementary school.

    c. parental involvement has a moderate effect on a child's educational achievement levels in elementary school.

    d. the more a parent is involved in their elementary school child's education the higher the child's achievement levels are likely to be.

19. The same researcher from question 18 found a correlation of .25 between parental involvement in education and high school students' overall achievement levels. This suggests that

    a. parents are less involved in their high school children's education.

    b. parental involvement has no effect on high school students' educational achievement levels.

    c. parental involvement is not as strongly related to educational achievement levels for high school students.

    d. the more a parent is involved in their high school student's education the lower the student's achievement levels.

20. A researcher investing the relationship between PSEO (Post Secondary Enrollment Options in which high school students can earn college credit) credits and third-year college dropout rates found a correlation of -.32. This correlation suggests that

    a. the more PSEO credits a student has the more likely they are to drop out of college by the third year.

    b. the more PSEO credits a student has the less likely they are to drop out of college by the third year.

    c. the more PSEO credits a student has the more likely they are to graduate college within four years.

    d. the fewer PSEO credits a study has the less likely they are to drop out of college by the third year.

EXERCISES

1. Compute the Pearson product-moment correlation coefficient for the following two variables:

Years of Education	Income (000)
8	12
9	15
11	14
12	20
14	28
14	33
16	33
16	45
18	48
22	61
23	81

2. You find that the correlation between two variables is equal to +.76. First, judge the sign (+ or –) and strength of the correlation. Is it direct or indirect? Now, calculate the coefficient of determination and the coefficient of alienation. What do both of these values mean?

3. Complete SPSS question 3 (below) and then interpret the score you obtained.

SHORT-ANSWER/ESSAY QUESTIONS

1. To determine the relationship between two continuous variables, you run a correlation in SPSS, expecting to find a sizable positive correlation between the variables. However, the SPSS output shows the Pearson's r is small. Should you give up on your hypothesis about the relationship between these variables? How else could you examine the data? What might a figure tell you about the relationship of your variables?

2. Correlations do not represent causality. Therefore, how could you describe the value of using a correlation to examine your data?

3. Three main types of correlation coefficients are described in this chapter (i.e., correlation coefficient, coefficient of determination, and coefficient of alienation). Please describe the identifying features of each.

SPSS QUESTIONS

1. Input the data given for question 1 of the "Exercises" section into SPSS. Now, calculate the correlation coefficient using SPSS. Does your result match what you calculated by hand?

2. Create a scatterplot in SPSS using the same data. What does it show?

3. Open the supplemental data set "Teacher Survey Data" in SPSS. Use SPSS to calculate the correlation coefficient for the relationship between "Overall Stress" and "Overall Satisfaction" levels in teachers.

JUST FOR FUN/CHALLENGE YOURSELF

1. Describe a curvilinear relationship. Come up with an example of two variables that may have a curvilinear relationship and explain why the relationship is curvilinear (use an example other than the one that was presented in the book).

2. If you are computing correlation coefficients among 10 variables, how many unique correlation coefficients will you calculate in total?

ANSWER KEY

TRUE/FALSE QUESTIONS

1. False. Even if two variables are correlated with each other, one variable does not necessarily cause the other. The example presented in this chapter's discussion of the correlation between ice cream consumption and crime illustrates this possibility very well.

2. True. Correlations are conducted on your entire set of data: This means that, while any correlations that you find are true for the entire data set, they're not necessarily true for each individual case or person. For example, if you found a direct (positive) correlation between years of education and income, this would not necessarily mean that all individuals with low education have low income (think Bill Gates). In practice, you will typically find some cases that diverge from the correlation.

3. False. A negative correlation, on its own, is neither any better nor any worse than a positive correlation.

4. False. While no correlation has been given one would assume that the more a student pays attention in class the higher an exam grade should be, suggesting a positive correlation.

5. True. This number is positive and therefore shows that the two variables increase together.

MULTIPLE-CHOICE QUESTIONS

1. (c) −1 to +1
2. (a) Two interval variables
3. (b) Indirect correlation
4. (a) direct
5. (c) −.70
6. (c) +.12
7. (a) *r*
8. (b) 0
9. (a) lower
10. (a) a direct
11. (a) .8 to 1.0
12. (c) .4 to .6
13. (c) −.98
14. (e) .0 to .2
15. (e) Phi coefficient
16. (c) Point biserial
17. (b) Spearman rank coefficient
18. (d) the more a parent is involved in their elementary school child's education the higher the child's achievement levels are likely to be. (c) is incorrect because of the word "effect," which suggests causation.
19. (c) parental involvement is not as strongly related to educational achievement levels for high school students.
20. (b) the more PSEO credits a student has the less likely they are to drop out of college by the third year.

EXERCISES

1.

	Years of Ed	Income	X²	Y²	XY
	8	12	64	144	96
	9	15	81	225	135
	11	14	121	196	154
	12	20	144	400	240
	14	28	196	784	392
	14	33	196	1,089	462
	16	33	256	1,089	528
	16	45	256	2,025	720
	18	48	324	2,304	864
	22	61	484	3,721	1,342
	23	81	529	6,561	1,863
Totals:	163	390	2,651	18,538	6,796

$$
\begin{aligned}
r_{xy} &= \frac{n\Sigma XY - \Sigma X \Sigma Y}{\sqrt{(n\Sigma X^2 - (\Sigma X)^2)(n\Sigma Y^2 - (\Sigma Y)^2)}} \\
&= \frac{11 \times 6{,}796 - 163 \times 390}{\sqrt{(11 \times 2{,}651 - 163^2)(11 \times 18{,}538 - 390^2)}} \\
&= \frac{74{,}756 - 63{,}570}{\sqrt{(2{,}592)(51{,}818)}} \\
&= .9652
\end{aligned}
$$

2. First, this correlation coefficient was found to be positive, meaning that this is a direct correlation. This means that the two variables included in the analysis tend to "move together"; in other words, as one variable increases, the other is expected to increase, and as one variable decreases, the other is expected to decrease. Next, as the correlation was found to be +.76, you could state that there is a strong relationship between these two variables.

   Next, the coefficient of determination is simply calculated as the square of the correlation coefficient. In this example, the coefficient of determination is equal to .58. The coefficient of alienation is equal to 1 minus the coefficient of determination, or 1 −.58 = .42. From the coefficient of determination, it can be said that 58% of the variance in one of the variables is explained by the variation in the second variable. The coefficient of alienation means that 42% of the variance in either of the variables cannot be explained by the other variable.

3. Completing SPSS question 3 should result in a correlation of −.352 for the relationship between teachers' overall stress and their overall satisfaction with the school where they teach. This number can be interpreted to mean that there is a weak, negative correlation between the two variables such that, the higher a teacher's stress the lower their overall satisfaction or visa versa, the higher their overall satisfaction the lower their stress. While the .35 is a weak

correlation this SPSS output shows it to be significant (something we get to in later chapters). Also, note that the variable you put first when interpreting a correlation does not matter as long as you have the direction correct and you say something about the strength or the significance.

SHORT-ANSWER/ESSAY QUESTIONS

1. No, the Pearson's *r* correlation does not tell the whole story. The next thing to look at is a scatterplot/scattergram. If the dots on the scatterplot are in no particular order, then you can accept the value of the correlation. If, however, the dots on the scatterplot show either a *U* shape or an inverted (upside-down) *U* shape, that would indicate a curvilinear relationship. This means that as one continuous variable changes, the relationship with another continuous variable either gets stronger, plateaus, and then gets weaker or the opposite—the relationship with the other continuous variable gets weaker, stays stable, and then gets stronger again.

2. Even if neither variable can be said to cause a change in the other, the correlation still conveys important messages. A correlation indicates the strength of a relationship between two continuous variables. It can further show a positive relationship (the variables change in the same direction) or a negative relationship (as one variable goes up, the other variable goes down). The coefficient of determination can also show the amount of variability in one variable that is accounted for by the variability in the other variable. All of this information can give the researcher ideas about how much of the change in a variable potentially remains to be described by other variables.

3. A correlation coefficient is a number—numerical index—that reflects the relationship between two variables. The coefficient of determination is the percentage of the variance in one variable that is accounted for by the variance in the other variable. The coefficient of alienation/nondetermination is the amount of variance in one variable left over and *not* accounted for by the variance in the other variable.

SPSS QUESTIONS

1. The following table presents the correct SPSS output. As you can see, the correlation coefficient matches the one calculated by hand.

Correlations

		Education	**Income (1,000s)**
Education	Pearson Correlation	1	.965**
	Sig. (2-tailed)		.000
	N	11	11
Income (1,000s)	Pearson Correlation	.965**	1
	Sig. (2-tailed)	.000	
	N	11	11

*** Correlation is significant at the 0.01 level (2-tailed).*

2. The following figure presents this scatterplot. Based on the scatterplot, there appears to be a direct, or positive, and quite strong relationship between these two variables.

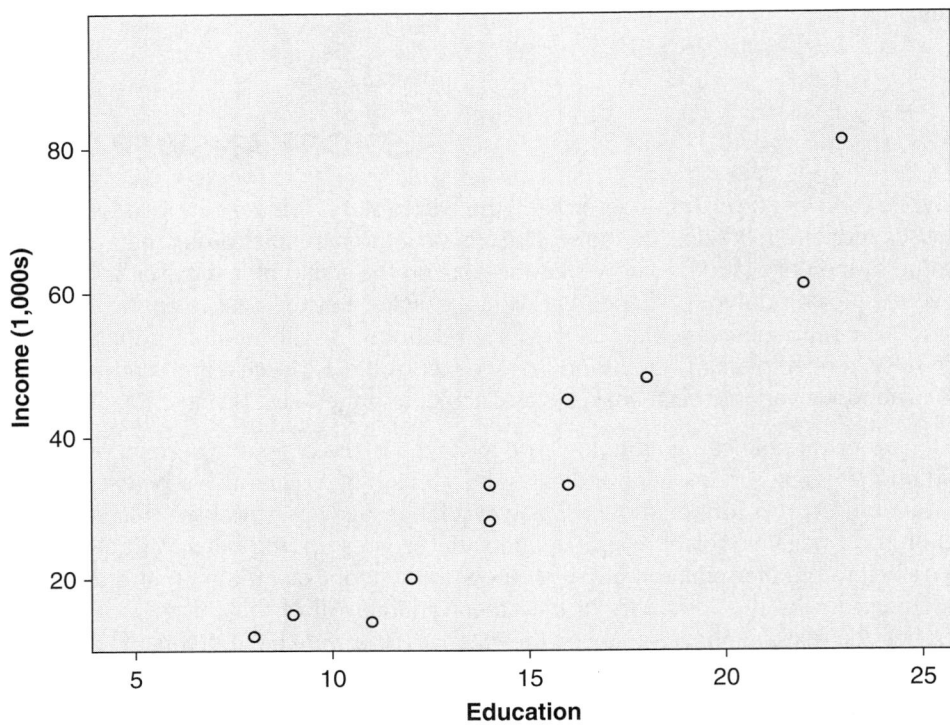

3. The following table presents the correct SPSS output.

Correlations

		Overall Satisfaction	Overall Stress
OverallSatisfaction	Pearson Correlation	1	−.352**
	Sig. (2-tailed)		.003
	N	70	70
OverallStress	Pearson Correlation	−.352**	1
	Sig. (2-tailed)	.003	
	N	70	70

** Correlation is significant at the 0.01 level (2-tailed).

JUST FOR FUN/CHALLENGE YOURSELF

1. In a curvilinear relationship, the nature of the relationship between two variables substantially changes over the range of these variables. Specifically, a direct, or positive, correlation between two variables becomes an indirect, or negative, relationship (or vice versa). There are many possible examples. One example could be the relationship between work satisfaction and hours worked per week. To illustrate, individuals who only work a few hours per

week may be very dissatisfied with their work, because their income is very low and they're not able to work as many hours as they would like. However, individuals who work very long hours— for example, 70 to 80 hours a week or more—may also be very dissatisfied with their work due to stress and a lack of free time. Individuals who work around a normal workweek, 40 hours per week, could have the greatest levels of job satisfaction, because they both make enough income and have enough free time. If this relationship were plotted on a graph, it would be in the shape of an upside-down U: Individuals with the fewest hours also have the lowest satisfaction. Then satisfaction increases as hours per week increase, up to around 40 hours per week. Beyond this point, satisfaction decreases as hours per week continue to increase. When hours per week are low, the relationship between these two variables is direct. However, when hours per week are high, the relationship between these two variables is indirect.

2. This can be calculated in the following way:

Number of unique correlations $\dfrac{n(n-1)}{2} \Rightarrow \dfrac{10(10-1)}{2} = 45.$

6 An Introduction to Understanding Reliability and Validity

Just the Truth

CHAPTER OUTLINE

- ✦ An Introduction to Reliability and Validity
  - ✧ What's Up With This Measurement Stuff?
- ✦ Reliability: Doing It Again Until You Get It Right
  - ✧ Test Scores: Truth or Dare?
  - ✧ Observed Score = True Score + Error Score
- ✦ Different Types of Reliability
  - ✧ Test-Retest Reliability
  - ✧ Parallel Forms Reliability
  - ✧ Internal Consistency Reliability
  - ✧ Interrater Reliability
- ✦ How Big Is Big? Finally: Interpreting Reliability Coefficients
  - ✧ And If You Can't Establish Reliability . . . Then What?
  - ✧ Just One More Thing
- ✦ Validity: Whoa! What Is the Truth?
  - ✧ Different Types of Validity
  - ✧ Content-Based Validity
  - ✧ Criterion-Based Validity
  - ✧ Construct-Based Validity
  - ✧ And If You Can't Establish Validity . . . Then What?
- ✦ A Last Friendly Word
- ✦ Validity and Reliability: Really Close Cousins

◆ Real-World Stats

◆ Summary

◆ Time to Practice

LEARNING OBJECTIVES

- Understand the difference between reliability and validity and learn why they are important.

- Learn the difference between the different types of reliability and validity.

- Understand what steps you can take if you need to increase reliability or validity.

- Learn how to compute and interpret different types of reliability and validity, both by hand and with SPSS.

SUMMARY/KEY POINTS

- Before we start analyzing and interpreting data, it will be important to understand what reliability and validity are, because they both have very important implications for the data.
  - Reliability explores this question: How do I know that the test, scale, instrument, etc., I use works every time I use it?
  - Validity explores this question: How do I know that the test, scale, instrument, etc., I use measures what it is supposed to?
  - If data are not reliable or not valid, then the results of any test or hypothesis will have to be inconclusive, as you are unsure about the quality of the data.

- A dependent variable is the outcome or predicted variable in an analysis, while an independent variable is the treatment or predictor variable in an analysis.

- Reliability relates to the degree to which a test measures something consistently.
  - Observed scores are the actual or measured scores, while true scores are the scores research participants would receive if the test contained no error.
    - The difference between these two scores is called the error score.
    - Observed scores may come close to true scores, but in the social and behavioral sciences, they are rarely the same as true scores due to the presence of error.
    - The less error you have, the greater the reliability.
  - Types of reliability include test-retest reliability, parallel forms reliability, internal consistency reliability, and interrater reliability.
    - Test-retest reliability is calculated as the correlation between scores from time 1 and scores from time 2.
    - Parallel forms reliability is calculated as the correlation between scores from the first form of your test with the scores from the second form of your test.
    - Internal consistency reliability is calculated using Cronbach's alpha.
    - Interrater reliability is calculated as the number of agreements between your two raters divided by the total number of possible agreements.
    - For high reliability, you want your reliability coefficients to be positive and to be as large as possible (1 is the highest possible reliability coefficient score).

○ If your test is not reliable, you must try to lower your error. A number of possible changes can be made for this purpose, including increasing the number of items or observations and deleting unclear items.

- Validity is the degree to which an assessment tool measures what it says it does.
  ○ Measures of validity include content validity, criterion validity, and construct validity.
    ○ Content validity is established by consultation with a content expert on the topic your instrument focuses on or for teachers making sure the tests they create adequately sample from the content taught. There is nothing worse (well, there are worse things) than a test in which half of the items are on something the teacher talked about for 5 minutes in class. Criterion validity is determined on the basis of the association between the test scores and some specified present or future criterion. Part of the validity of college entrance exams (e.g., ACT, SAT) is that they positively correlate with attainment of a college degree.
    ○ Construct validity is based on a judgment of how well your test reflects an underlying construct or idea.
  ○ If a test or instrument lacks validity, changes can be made to improve validity. What changes are appropriate to make will depend on the type of validity in question.
    ○ Content validity: Redo questions so that, according to an expert judge, they more accurately reflect the topics and ideas being tested for. As a teacher, make sure your test samples from all of the content taught (the entire unit or chapter the test is covering).
    ○ Criterion validity: Reexamine the nature of the items on the test and question whether responses should be expected to relate to the criterion you selected.
    ○ Construct validity: Review the theoretical rationale that underlies your test.
  ○ When working on a thesis or dissertation, it is strongly recommended that an instrument or text be used that has already been established to be reliable and valid.
  ○ A test can be reliable and not valid, but it is impossible to have a valid test that is not reliable.
    ○ The maximum level of validity possible is equal to the square root of the reliability coefficient.
    ○ Most state and nationally standardized tests used in schools and colleges today (ACT, ITBS, TerraNova, CAT, etc.) have extensive reliability and validity studies completed on them and information on these studies is available in the test manual.

KEY TERMS

- **Dependent variable**: The outcome of or predicted variable in an analysis

- **Independent variable**: A treatment variable that is manipulated or the predictor variable in an analysis

- **Reliability**: The degree to which a test measures something consistently
  ○ **Observed score**: The actual score that is recorded or observed
  ○ **True score**: The score that you would receive if a test measured your ability perfectly
  ○ **Error score**: The part of a test score that is random and contributes to the unreliability of a test

○ **Test-retest reliability**: A type of reliability that examines consistency over time
○ **Parallel forms reliability**: A type of reliability that examines consistency across different forms of the same test
○ **Internal consistency reliability**: A type of reliability that measures the extent to which items on a test are consistent with one another and represent one—and only one—dimension, construct, or area of interest
○ **Cronbach's alpha**: A particular measure of internal consistency reliability
○ **Interrater reliability**: A type of reliability that examines the consistency of raters

• **Validity**: The quality of a test such that it measures what it says it does
○ **Content validity**: A type of validity that examines the extent to which a test accurately reflects all possible topics and ideas under the subject or topic you are testing for
○ **Criterion validity**: A type of validity that examines how well a test reflects some criterion that occurs in the present or future
○ **Concurrent criterion validity**: A type of validity that examines how well a test outcome is consistent with a criterion that occurs in the present
○ **Predictive validity**: A type of validity that examines how well a test outcome is consistent with a criterion that occurs in the future
○ **Construct validity**: A type of validity that examines how well a test reflects an underlying construct or idea

TRUE/FALSE QUESTIONS

1. Reliability and validity should be determined after an analysis is already complete.

2. If a test or instrument contains no error, then it can be said to have perfect reliability.

3. Observed scores are commonly exactly the same as true scores.

4. Smaller amounts of error are associated with greater reliability.

5. High reliability is associated with small reliability coefficients, as close as possible to 21.

6. If reliability is very low, nothing needs to be done before reporting this finding in your paper.

7. When working on your thesis or dissertation, it is best to create your own test instrument.

8. Teachers don't need to worry about the validity or reliability of the tests they create for their students.

9. Teachers should use test-retest reliability to determine if their tests are reliable.

MULTIPLE-CHOICE QUESTIONS

1. The manipulated treatment, or predictor variable in an analysis, is known as _____ variable.

    a. the dependent

    b. the independent

    c. the correlated

    d. a skewed

2. The outcome, or predicted variable in an analysis, is known as _____ variable.

    a. the dependent

    b. the independent

    c. the correlated

    d. a skewed

3. Reliability serves to answer which of the following questions?

    a. How do I know that the test, scale, instrument, etc., I use measures what it is supposed to?

    b. How do I know that the test, scale, instrument, etc., I use works on all populations?

    c. How do I know that the test, scale, instrument, etc., I use works every time I use it?

4. Validity serves to answer which of the following questions?

    a. How do I know that the test, scale, instrument, etc., I use works every time I use it?

    b. How do I know that the test, scale, instrument, etc., I use works on all populations?

    c. How do I know that the test, scale, instrument, etc., I use measures what it is supposed to?

5. The actual, or measured, score a student received on a test is called the _____ score.

    a. true

    b. observed

    c. measured

    d. error

    e. perfect

6. The score that you would receive if a test contained no error is called the _____ score.

    a. true

    b. observed

    c. measured

    d. error

    e. perfect

7. The difference between the score received if a test contained no error and the actual/measured score is called the _____ score.

   a. true

   b. observed

   c. measured

   d. error

   e. perfect

8. Test-retest reliability examines which of the following?

   a. The consistency of raters

   b. The consistency across different forms of the same test

   c. The extent to which items in a test are consistent with one another and represent exactly one dimension, construct, or area of interest

   d. Consistency over time

9. Parallel forms reliability examines which of the following?

   a. The consistency of raters

   b. The consistency across different forms of the same test

   c. The extent to which items in a test are consistent with one another and represent exactly one dimension, construct, or area of interest

   d. Consistency over time

10. Internal consistency reliability examines which of the following?

    a. The consistency of raters

    b. The consistency across different forms of the same test

    c. The extent to which items in a test are consistent with one another and represent exactly one dimension, construct, or area of interest

    d. Consistency over time

11. Interrater reliability examines which of the following?

    a. The consistency of judges

    b. The consistency across different forms of the same test

    c. The extent to which items in a test are consistent with one another and represent exactly one dimension, construct, or area of interest

    d. Consistency over time

12. _____ reliability is calculated using Cronbach's alpha.

    a. Test-retest

    b. Parallel forms

    c. Internal consistency

    d. Interrater

13. _____ reliability is calculated as the number of agreements between judges divided by the total number of possible agreements.

   a. Test-retest

   b. Parallel forms

   c. Internal consistency

   d. Interrater

14. _____ reliability is calculated as the correlation between scores from time 1 and scores from time 2.

   a. Test-retest

   b. Parallel forms

   c. Internal consistency

   d. Interrater

15. _____ reliability is calculated as the correlation between scores from the first form of the test with the scores from the second form of the test.

   a. Test-retest

   b. Parallel forms

   c. Internal consistency

   d. Interrater

16. _____ validity examines how well a test reflects some standard that occurs in the present or future.

   a. Predictive

   b. Concurrent criterion

   c. Criterion

   d. Construct

   e. Content

17. _____ validity examines how well a test reflects an underlying idea.

   a. Predictive

   b. Concurrent criterion

   c. Criterion

   d. Construct

   e. Content

18. _____ validity examines the extent to which a test accurately reflects all possible topics and ideas under the subject or topic it is designed to test for.

   a. Predictive

   b. Concurrent criterion

   c. Criterion

　　d. Construct

　　e. Content

19. If you need to improve content validity, you should do which of the following?

　　a. Reexamine the nature of the items on the test and question whether responses should be expected to relate to the criterion you selected.

　　b. Review the theoretical rationale that underlies the test.

　　c. Redo questions so that, according to an expert judge, they more accurately reflect the topics and ideas being tested for.

20. If you need to improve criterion validity, you should do which of the following?

　　a. Reexamine the nature of the items on the test and question whether responses should be expected to relate to the criterion you selected.

　　b. Review the theoretical rationale that underlies the test.

　　c. Redo questions so that, according to an expert judge, they more accurately reflect the topics and ideas being tested for.

21. If you need to improve construct validity, you should do which of the following?

　　a. Reexamine the nature of the items on the test and question whether responses should be expected to relate to the criterion you selected.

　　b. Review the theoretical rationale that underlies the test.

　　c. Redo questions so that, according to an expert judge, they more accurately reflect the topics and ideas being tested for.

22. Which of the following is a correct statement?

　　a. The maximum reliability is equal to the square root of the validity.

　　b. The maximum validity is equal to the square root of the reliability.

　　c. The maximum reliability is equal to the square of the validity.

　　d. The maximum validity is equal to the square of the reliability.

EXERCISES

1. Say that you were very tired when taking an exam and did much worse than you had anticipated, getting a score of 72. If you had been wide awake and energetic, and if the test perfectly measured your knowledge, you would have gotten a score of 94. What is your error score?

2. The following data illustrate a set of scores from an instrument administered to eight pre-service teachers at two different times. Calculate the test-retest reliability.

ID	Time 1	Time 2
1	72	76
2	54	43
3	86	92
4	92	94
5	99	87
6	43	69
7	76	43
8	84	92

3. The following data illustrate a set of scores from a test administered to English-language learners in two different forms to five respondents. Calculate the parallel forms reliability.

ID	Form 1	Form 2
1	8.7	92
2	7.6	7.9
3	3.4	4.3
4	3.7	5.3
5	5.4	6.1

4. You and your friend Robert join a film club and, over the course of a month, watch 10 films together. While you have similar taste in films, you don't agree on every movie you see. The following data illustrate whether you and Robert believe these films to be good (G) or bad (B). Calculate the interrater reliability.

Film	1	2	3	4	5	6	7	8	9	10
You	G	B	G	G	G	B	G	G	B	G
Robert	G	G	B	G	G	B	G	G	B	B

SHORT-ANSWER/ESSAY QUESTIONS

1. Explain the difference between test-retest reliability and parallel forms reliability.

2. Explain why teachers should be concerned about the reliability of the classroom tests they create.

SPSS QUESTIONS

1. Input the data from question 1 under the "Just for Fun/Challenge Yourself" section below into SPSS and calculate Cronbach's alpha. If you complete this "Challenge Yourself" question manually, do the two figures match?

2. Open the supplemental data set "Teacher Survey Data" in SPSS. Use SPSS to calculate test-retest reliability for SurveyItemA12 and SurveyItemB12, which both state "I am generally satisfied with being a teacher at this school."

JUST FOR FUN/CHALLENGE YOURSELF

1. The following data present scores for 10 individuals on a 5-item test. Using these data, calculate Cronbach's alpha. Now, do a little additional research on this statistic. Is the score you calculated good or bad? Is it high enough to be considered acceptable?

ID	Item 1	Item 2	Item 3	Item 4	Item 5
1	8	9	7	8	8
2	5	6	8	4	5
3	1	2	1	1	1
4	4	3	4	5	4
5	1	3	2	3	2
6	5	4	5	6	5
7	2	3	4	2	3
8	9	7	8	7	9
9	6	5	8	5	6
10	4	5	5	4	5

ANSWER KEY

TRUE/FALSE QUESTIONS

1. False. Reliability and validity should always be examined before an analysis is conducted. If the reliability and/or validity is not up to par, you risk having inconclusive results.

2. True. Error is measured as the difference between the true score and the observed (measured) score. If these two scores were found to be exactly the same, you would have no error, or perfect reliability.

3. False. This is extremely rare in the social and behavioral sciences due to the common presence of error.

4. True. The less error you have, the greater your reliability.

5. False. High reliability is in fact associated with high reliability coefficients, as close as possible to 1.

6. False. Having very low reliability is a serious concern. The steps appropriate for the type of reliability in question should be taken to attempt to increase the level of reliability.

7. False. Instead, it is best to use a previously published test instrument whose reliability and validity have already been determined to be at least adequate. Otherwise, you run the serious and substantial risk of having your test instrument lack sufficient reliability and/or validity, which would call your data into serious question and force your results to be inconclusive.

8. False. If you want to get a true picture of what your students know you need to create tests that are both reliable and valid.

9. False. This would be a waste of valuable classroom time and your students would likely do better on the second test given that they have had the test once already. A better method would be a measure of internal consistency.

MULTIPLE-CHOICE QUESTIONS

1. (b) the independent

2. (a) the dependent

3. (c) How do I know that the test, scale, instrument, etc., I use works every time I use it?

4. (c) How do I know that the test, scale, instrument, etc., I use measures what it is supposed to?

5. (b) observed

6. (a) true

7. (d) error

8. (d) Consistency over time

9. (b) The consistency across different forms of the same test

10. (c) The extent to which items in a test are consistent with one another and represent exactly one dimension, construct, or area of interest

11. (a) The consistency of judges

12. (c) Internal consistency

13. (d) Interrater

14. (a) Test-retest

15. (b) Parallel forms

16. (c) Criterion

17. (d) Construct

18. (e) Content

19. (c) Redo questions so that, according to an expert judge, they more accurately reflect the topics and ideas being tested for.

20. (a) Reexamine the nature of the items on the test and question whether responses should be expected to relate to the criterion you selected.

21. (b) Review the theoretical rationale that underlies the test you developed.

22. (b) The maximum validity is equal to the square root of the reliability.

1. Error score = Observed score − True score = 72 − 94 = −22.

2. The necessary calculations for this question are shown here. In essence, this question necessitates calculation of the correlation between time 1 scores and time 2 scores.

ID	Time 1	Time 2	Time 1²	Time 2²	Time 1 * Time 2
1	72	76	5,184	5,776	5,472
2	54	43	2,916	1,849	2,322
3	86	92	7,396	8,464	7,912
4	92	94	8,464	8,836	8,648
5	99	87	9,801	7,569	8,613
6	43	69	1,849	4,761	2,967
7	76	43	5,776	1,849	3,268
8	84	92	7,056	8,464	7,728
Sum:	606	596	48,442	47,568	46,930

$$r_{xy} = \frac{n\Sigma XY - \Sigma X \Sigma Y}{\sqrt{(n\Sigma X^2 - (\Sigma X)^2)(n\Sigma Y^2 - (\Sigma Y)^2)}}$$

$$= \frac{8 \times 46,930 - 606 \times 596}{\sqrt{(8 \times 48,442 - 606^2)(8 \times 47,568 - 596^2)}}$$

$$= \frac{375,440 - 361,176}{\sqrt{(20,300)(25,328)}}$$

$$= .6291$$

3. The following illustrates the calculations necessary to calculate parallel forms reliability. In essence, this question requires the calculation of the correlation between form 1 scores and form 2 scores.

ID	Form 1	Form 2	Form 1²	Form 2²	Form 1 * Form 2
1	8.7	9.2	75.69	84.64	80.04
2	7.6	7.9	57.76	62.41	60.04
3	3.4	4.3	11.56	18.49	14.62
4	3.7	5.3	7.29	28.09	14.31
5	5.4	6.1	29.16	37.21	32.94
Sum:	27.8	32.8	181.46	230.84	201.95

$$r_{xy} = \frac{n\Sigma XY - \Sigma X \Sigma Y}{\sqrt{(n\Sigma X^2 - (\Sigma X)^2)(n\Sigma Y^2 - (\Sigma Y)^2)}}$$

$$= \frac{5 \times 201.95 - 27.8 \times 32.8}{\sqrt{(5 \times 181.46 - 27.8^2)(5 \times 230.84 - 32.8^2)}}$$

$$= \frac{1{,}009.75 - 911.84}{\sqrt{(134.46)(78.36)}}$$

$$= .9539$$

4. The following illustrates the calculation of the interrater reliability for this question: Divide the number of total agreements by the number of total possible agreements.

Film	1	2	3	4	5	6	7	8	9	10
You	G	B	G	G	G	B	G	G	B	G
Robert	G	G	B	G	G	B	G	G	B	B

$$\text{Interrater reliability} = \frac{n \text{ Agreements}}{n \text{ Possible agreements}} = \frac{7}{10} = .7$$

SHORT-ANSWER/ESSAY QUESTIONS

1. Test-retest reliability measures reliability, or consistency over time, while parallel forms reliability examines consistency across multiple forms of the same test. Test-retest reliability would be calculated by determining the correlation between a test given at one time and the same test given at a second time, while parallel forms reliability would be calculated by determining the correlation between administrations of two versions of the test instrument.

2. If a test created by a teacher is not reliable it means that the scores students receive on the test have a large amount of error in them and therefore do not reflect the students' true knowledge of the content.

SPSS QUESTIONS

1. The SPSS output for this analysis is presented below. The calculated Cronbach's alpha score is 0.969, which is identical to the figure you would have arrived at by calculating manually.

Case Processing Summary

		N	%
Cases	Valid	10	100.0
	Excluded[a]	0	.0
	Total	10	100.0

[a] Listwise deletion based on all variables in the procedure.

Reliability Statistics

Cronbach's Alpha	N of Items
.969	5

2. The SPSS output for this analysis is presented below. To calculate the test–retest reliability for these two items you simply perform a Pearson's r. A Pearson's r calculation will result in a correlation of .783 between the first time teachers answered this survey question (at midyear) and the second time (end of the year).

Correlations

		I am generally satisfied with being a teacher at this school. (midyear)	I am generally satisfied with being a teacher at this school. (end of year)
I am generally satisfied with being a teacher at this school. (midyear)	Pearson Correlation	1	.783**
	Sig. (2-tailed)		.000
	N	70	70
I am generally satisfied with being a teacher at this school. (last week of school)	Pearson Correlation	.783**	1
	Sig. (2-tailed)	.000	
	N	70	70

** Correlation is significant at the 0.01 level (2-tailed).

JUST FOR FUN/CHALLENGE YOURSELF

1. First, the variances for all five items as well as the variance for total score need to be calculated. As this calculation was covered in Chapter 3, it is not repeated here. The final column of the following table gives the variances of each of the five individual items as well of the total score.

ID	Item 1	Item 2	Item 3	Item 4	Item 5	Total
1	8	9	7	8	8	40
2	5	6	8	4	5	28
3	1	2	1	1	1	6
4	4	3	4	5	4	20
5	1	3	2	3	2	´1
6	5	4	5	6	5	25
7	2	3	4	2	3	14
8	9	7	8	7	9	40
9	6	5	8	5	6	30
10	4	5	5	4	5	23
Variance:	7.39	4.68	6.40	4.72	6.18	130.46

Next, the variances for the five individual items need to be summed:

Sum of item variances = 29.37.

Finally, the following equation presents the calculation for Cronbach's alpha:

$$\alpha = \left(\frac{k}{k-1}\right)\left(\frac{s_Y^2 - \Sigma s_i^2}{s_Y^2}\right)$$

$$\Rightarrow \left(\frac{5}{5-1}\right)\left(\frac{130.46 - 29.37}{130.46}\right)$$

$$= .9686$$

These five items have a Cronbach's alpha score of 0.9686. Generally, alpha scores of 0.7 or higher are considered acceptable. Thus, these five items have a very high and very acceptable alpha score.

7 Hypotheticals and You

Testing Your Questions

LEARNING OBJECTIVES

- Learn the differences between a sample and a population and how each is related to your research.

- Understand the difference between null and research hypotheses, as well as directional and nondirectional hypotheses.

- Learn how to create good hypotheses.

SUMMARY/KEY POINTS

- A hypothesis is an "educated guess" that describes the relationship between variables. In essence, it is like a more specific, directly testable version of a research question.

- In a study, a sample is drawn from a larger population, and analyses are conducted on the sample. Optimally, it is possible to generalize your results to the population.
  - Hypothesis testing relates to the sample itself, not the population.
  - A *representative* sample must be used if you wish to generalize the results of your analyses to the population at large. Individuals in a representative sample should match as closely as possible to the characteristics of the population (and the study must also meet more specific methodological requirements).

- Sampling error is how well a sample approximates the characteristics of a population.
  - Higher sampling error means a greater difference between the sample statistic and the population parameter, so it is more difficult to generalize your results to the population.

- The two types of hypotheses are the null and alternative hypotheses.
  - The null hypothesis is formulated first and states that there is no relationship between your variables.
  - The research hypothesis, formulated second, states that there is a relationship between your variables.
  - A research hypothesis that suggests the direction of the relationship is called a *directional* hypothesis. One-tailed tests can be used with these hypotheses.
  - A research hypothesis that does not suggest the direction of the relationship is called a nondirectional hypothesis. Two-tailed tests should be used with these hypotheses.
  - Unless you have sufficient evidence otherwise, you must assume that the null hypothesis is true.
  - Null hypotheses always refer to the population, while research hypotheses always refer to the sample. Therefore, the null hypothesis is only indirectly tested (making it an *implied* hypothesis), while the research hypothesis can be tested directly.
  - While null hypotheses are written using Greek symbols, research hypotheses are written using Roman symbols (letters from the English alphabet).
  - Good hypotheses have the following features:
    - They are stated in declarative form, not as a question.
    - They posit an expected relationship between variables.
    - They reflect the theory or literature on which they are based.
    - They are brief and to the point.
    - They are testable.

KEY TERMS

- **Hypothesis**: An "educated guess" describing the relationship between two or more variables

- **Population**: A larger group of respondents from whom a sample is collected and to which you hope to generalize after conducting your analyses

- **Sample**: A subset taken from a population for the purposes of your study. Data are collected on a sample, and, optimally, the results are generalized to the population

- **Sampling error**: The difference between sample and population values

- **Null hypothesis**: A statement of equality between sets of variables

- **Research hypothesis**: A statement that there is a relationship between variables

- **Nondirectional research hypothesis**: A hypothesis that reflects a difference between groups but does not specify the direction of the difference

- **Directional research hypothesis**: A hypothesis that reflects a difference between groups and also specifies the direction of the difference

- **One-tailed test**: A directional test, which reflects a directional hypothesis

- **Two-tailed test**: A nondirectional test, which reflects a nondirectional hypothesis

TRUE/FALSE QUESTIONS

1. A research question is a more specific, testable version of a hypothesis.

2. The null hypothesis is only indirectly tested, making it an implied hypothesis.

3. Good hypotheses should be (along with other attributes) brief and to the point, testable, and stated in declarative form.

MULTIPLE-CHOICE QUESTIONS

1. When conducting a study, you draw a smaller _____ from a larger _____.

   a. population; sample

   b. sample; population

   c. null hypothesis; research hypothesis

   d. nondirectional hypothesis; directional hypothesis

2. To generalize your results, you need to have a _____.

   a. representative sample

   b. representative population

   c. population larger than your sample

   d. null hypothesis

3. A representative sample should have which of the following?

   a. A null hypothesis

   b. A nondirectional hypothesis

   c. A population

   d. A small level of sampling error

4. A high level of sampling error means that _____.

   a. the population may be too small

   b. your sample may be too large

   c. you may not be able to generalize to your population

   d. your hypotheses will not be supported

5. Which of the following types of hypotheses states that there is no relationship between your variables?

   a. The null hypothesis

   b. The research hypothesis

   c. The population hypothesis

   d. The sample hypothesis

6. Which of the following types of hypotheses states that there is a relationship between your variables?

   a. The null hypothesis

   b. The research hypothesis

   c. The population hypothesis

   d. The sample hypothesis

7. A one-tailed test would be used with a _____ hypothesis.

   a. null

   b. research

   c. directional

   d. nondirectional

8. A two-tailed test would be used with a _____ hypothesis.

   a. null

   b. research

   c. directional

   d. nondirectional

9. If you are unsure whether the null or research hypothesis is true, you must assume that _____.

   a. the population was too small

   b. the sample was too small

   c. the research hypothesis is true

   d. the null hypothesis is true

10. Null hypotheses always refer to _____.

    a. the sample

    b. the population

    c. sampling error

    d. two-tailed tests

11. Research hypotheses always refer to _____.

    a. the sample

    b. the population

    c. sampling error

    d. one-tailed tests

12. Null hypotheses are written using which of the following types of letters?

    a. Greek

    b. Roman

    c. Arabic

    d. Sanskrit

13. Research hypotheses are written using which of the following types of letters?

    a. Greek

    b. Roman

    c. Arabic

    d. Sanskrit

14. Which of the following is not a feature of good hypotheses?

    a. There should be no more than one null and one research hypothesis in any study.

    b. They should be brief and to the point.

    c. They should be testable.

    d. They should posit an expected relationship between variables.

15. Which type of hypothesis is this? *There is no relationship between socioeconomic status and drug use in high schoolers.*

    a. Research hypothesis

    b. Null hypothesis

    c. Sample hypothesis

    d. Directional research hypothesis

16. Which type of hypothesis is this? *Individuals with a college degree will have higher incomes than those with no college degree.*

    a. Null hypothesis

    b. Directional research hypothesis

    c. Nondirectional research hypothesis

    d. Sample hypothesis

17. Which type of hypothesis is this? *Johnson Middle School will differ from Rogers Middle School with regard to test scores.*

    a. Null hypothesis

    b. Directional research hypothesis

    c. Nondirectional research hypothesis

    d. Sample hypothesis

18. What type of hypothesis is this an example of? $H_1 : \overline{X}_A \neq \overline{X}_B$.

    a. Null hypothesis

    b. Directional research hypothesis

    c. Nondirectional research hypothesis

    d. Population hypothesis

19. What type of hypothesis is this an example of? $H_1 : \overline{X}_A < \overline{X}_B$.

    a. Null hypothesis

    b. Directional research hypothesis

    c. Nondirectional research hypothesis

    d. Population hypothesis

20. What type of hypothesis is this an example of? $H_0 : \mu_A = \mu_B$.

    a. Null hypothesis

    b. Directional research hypothesis

    c. Nondirectional research hypothesis

    d. Population hypothesis

SHORT-ANSWER/ESSAY QUESTIONS

1. You are going to conduct a study on the relationship between watching violence on television and violent behavior in your middle school. Generate a null hypothesis and two research hypotheses (nondirectional and directional) for this study.

2. Now, write both the null and research hypotheses using the appropriate Greek or Roman letters.

3. Check the hypotheses you generated against the five features of a good hypothesis. Do your hypotheses fulfill all of the requirements?

4. Come up with written descriptions of the hypotheses given under questions 18–20 in the "Multiple-Choice" section.

5. What is wrong with this hypothesis? *Will the accelerated class score higher on a reading comprehension exam as compared with a regular class?*

6. What is wrong with this hypothesis? *Today's district administrators are more concerned with the nutritional value of school lunches than ever before.*

JUST FOR FUN/CHALLENGE YOURSELF

Do some additional research on the difference between one-way and two-way statistical tests.

1. How do they differ?

2. Which of the two is more likely to be found significant, in general?

3. Why is it justified to use a one-way statistical test with a directional hypothesis but not with a nondirectional hypothesis?

ANSWER KEY

TRUE/FALSE QUESTIONS

1. False. The opposite is true—a hypothesis is a more specific, testable version of a research question.

2. True.

3. True. These are some of the characteristics of good hypotheses.

MULTIPLE-CHOICE QUESTIONS

1. (b) sample; population

2. (a) representative sample

3. (d) A small level of sampling error

4. (c) you may not be able to generalize to your population

5. (a) The null hypothesis

6. (b) The research hypothesis

7. (c) directional

8. (d) nondirectional

9. (d) the null hypothesis is true

10. (b) the population

11. (a) the sample

12. (a) Greek

13. (b) Roman

14. (a) There should be no more than one null and one research hypothesis in any study

15. (b) Null hypothesis

16. (b) Directional research hypothesis

17. (c) Nondirectional research hypothesis

18. (c) Nondirectional research hypothesis

19. (b) Directional research hypothesis

20. (a) Null hypothesis

SHORT-ANSWER/ESSAY QUESTIONS

1. An example of a null hypothesis: *With regard to the number of instances of violent behavior, there will be no relationship between individuals who watch less than 1 hour of violent television per day and those who watch 1 hour or more of violent television per day.*

    An example of a nondirectional research hypothesis: *With regard to the number of instances of violent behavior, there is a difference between individuals who watch less than 1 hour of violent television per day and those who watch 1 hour or more of violent television per day.*

    An example of a directional research hypothesis: *Individuals who watch 1 hour of violent television per day or more will exhibit a greater number of instances of violent behavior in school than those who watch less than 1 hour of violent television per day.*

2. Here are some examples (your subscripts do not have to match exactly):

    Null hypothesis: $H_0 : \mu_{<1hr} = \mu_{1+hr}$.

    Nondirectional research hypothesis: $H_1 : \bar{X}_{<1hr} \neq \bar{X}_{1+hr}$.

    Directional research hypothesis: $H_1 : \bar{X}_{<1hr} < \bar{X}_{1+hr}$.

3. As a reminder, the five features are these:
    ○ They are stated in declarative form, not as a question.
    ○ They posit an expected relationship between variables.
    ○ They reflect the theory or literature on which they are based.
    ○ They are brief and to the point.
    ○ They are testable.

4.

    a. For question 18: $H_1 : \bar{X}_A \neq \bar{X}_B$.

        The average score of individuals in Group A is different from the average score of individuals in Group B.

        For question 19: $H_1 : \bar{X}_A < \bar{X}_B$.

        Individuals in Group B will have higher scores, on average, than will individuals in Group A.

        For question 20: $H_0 : \mu_A = \mu_B$.

        There is no difference between the average score of individuals in Group A and the average score of individuals in Group B.

5. The problem with this hypothesis is that it is phrased as a question.

6. The problem with this hypothesis is that it is not testable (we don't have the data and can't collect it).

JUST FOR FUN/CHALLENGE YOURSELF

1. As mentioned in this chapter, one-way statistical tests are suited to directional hypotheses, while two-way tests are suited to nondirectional hypotheses. In essence, a two-way test examines both possibilities (i.e., that group 1 has a higher average than group 2 and that group 2 has a higher average than group 1). A one-way test examines only one possibility (i.e., that either group 1 has a higher average than group 2 or that group 2 has a higher average than group 1).

2. Because the one-way test examines only one possibility, it is easier for this test, if indeed your research hypothesis is true, to be found significant and for your research hypothesis to be adequately supported. Because a two-way test needs to examine both "sides," or both possibilities, it is weaker in a sense, and it is more difficult for you to find a significant result.

3. It is not justified to use a one-way test with a nondirectional hypothesis, as you are going into the analysis not knowing what you may find. Therefore, using a more powerful one-way test in situations where you don't initially start with a directional hypothesis is, in a sense, "cheating."

8 Probability and Why It Counts

Fun With a Bell-Shaped Curve

LEARNING OBJECTIVES

- Review the importance of probability in statistics.

- Understand the normal curve and its relation to the field of statistics.

- Learn how to compute z scores by hand and using SPSS.

SUMMARY/KEY POINTS

- The study of probability is the basis for the normal curve and the foundation for inferential statistics.
  - The normal curve provides a basis for understanding the probability associated with any possible outcome, such as attaining a certain score.
  - The study of probability is the basis for determining the degree of confidence we have in stating that a particular finding or outcome is true.
  - Probability allows us to determine the exact mathematical likelihood that a difference between groups, or an association between variables, is due to a practice or treatment rather than to chance or error.

- The normal curve is the basis for probability and statistics.
  - The normal curve has no skew and is perfectly symmetrical about the mean.
  - The tails of the normal distribution are asymptotic, meaning they never touch the horizontal axis, which is equal to zero.
  - In the social and behavioral sciences, as well as in other fields, many things are normally distributed, including measures such as height and IQ.
  - Events that occur in the extremes of the normal curve have a very small probability, while more "average" values are much more common.

- The normal curve has many specific statistical features.
  - Over 99.5% of scores are within 3 standard deviations of the mean.
  - Approximately 68% of scores fall within 1 standard deviation of the mean.
  - Exactly 50% of scores fall on either side of the distribution (i.e., either side of the mean).
  - The percentages or areas under the normal curve can be interpreted as probabilities.

- Standard scores are raw scores that have been adjusted for the particular mean and standard deviation of the distribution from which they are derived. They can be used to compare raw scores between different samples that have different distributions.

The most commonly used standard score is the z score.
  - To calculate the z score, you subtract the mean from the raw score and divide this difference by the standard deviation.
  - Scores that fall below the mean have negative z scores, while scores that fall above the mean have positive z scores.
  - The score located 1 standard deviation above the mean is "1 z score" above the mean.
  - We can use z scores and the normal distribution to determine the probability of some event occurring.
  - Skewness and kurtosis are measures that are used to describe the shape of a distribution.

- A statistical test can be used to determine the probability of the differences between groups or relationships between variables in the data. After the test is conducted, the calculated probability can be compared with a standard to see whether the result is "significant."

The standard of 5%, which is equivalent to a probability of .05, is the most commonly used standard in statistics. This means we need to be at least 95% sure of the difference between groups or the relationship between variables in order to call it "significant." This also means that a result is significant if we find a z score that has less than a 5% chance of occurring.

KEY TERMS

- **Normal curve** (bell-shaped curve): A distribution of scores that is symmetrical about the mean and in which the median, mean, and mode are all equal. This type of distribution has asymptotic tails, which never reach zero.

- **Asymptotic**: The quality of the normal curve such that its tails never touch the horizontal axis (equal to zero)

- **Standard scores**: Raw scores that are adjusted for the mean and standard deviation of the distribution from which they come

- **Standardized scores**: A score that comes from a distribution with a predefined mean and standard deviation

- **z score**: A specific type of standard score in which the mean of scores is subtracted from the raw score and then this difference is divided by the standard deviation
  - **Skewness**: A measure of the lack of symmetry, or "lopsidedness," of a distribution. A distribution that is skewed has one tail that is longer than the other.
    - **Positive skew**: A distribution that has many data points to the left and a long tail to the right
    - **Negative skew**: A distribution that has many data points to the right and a long tail to the left
  - **Kurtosis**: A measure that relates to how flat or peaked a distribution appears
    - **Platykurtic**: A distribution that is relatively flat compared to a normal, or bell-shaped, distribution
    - **Leptokurtic**: A distribution that is relatively peaked compared to a normal, or bell-shaped, distribution

TRUE/FALSE QUESTIONS

1. The percentages of scores under sections of the normal curve depend on the mean and standard deviation of distribution.

2. You can compare z scores across two or more different distributions.

3. Values for the area under the normal curve can be viewed/interpreted as probabilities.

4. When looking up z scores using a z table, it is very important to consider whether the z score you are looking up is positive or negative.

5. A standard score is the same as a standardized score; an example is the score you received on the SAT or ACT.

6. Skewness is a measure of the central point of a class interval.

7. Normal distributions are not common in education as students vary too much.

MULTIPLE-CHOICE QUESTIONS

1. Which of the following percentages of the normal curve reflects all scores greater than zero?

   a. 10%

   b. 25%

   c. 50%

   d. 100%

2. The entire normal curve represents _____ of scores.

   a. 25%

   b. 50%

   c. 99%

   d. 100%

3. With regard to the normal curve, _____ of scores are within 1 standard deviation of the mean.

   a. 13.59%

   b. 2.15%

   c. 68.26%

   d. 99.99%

4. Which of the following percentages of scores is within 2 standard deviations of the mean?

   a. 0.13%

   b. 2.15%

   c. 68.26%

   d. 95.44%

5. If your set of scores has a mean of 57, what is the z score for a raw score of 57?

   a. −1

   b. 0

   c. 1

   d. 2

6. If a raw score is above the mean, the z score must be _____.

   a. negative

   b. positive

    c. equal to zero

    d. impossible to compute

7. If a score is 4 standard deviations above the mean, then its z score must be equal to _____.

    a. – 4

    b. 4

    c. 0

    d. 1

    e. 4^4

8. Based on the normal curve, what percentage of scores have a z score of 2 or greater?

    a. 7.16%

    b. 22.42%

    c. 2.28%

    d. 0.15%

9. Based on the normal curve, what percentage of scores have a z score less than –0.5?

    a. 30.85%

    b. 22.42%

    c. 47.40%

    d. 23.65%

10. On your last exam, the class scored an average of 82 with a standard deviation of 8. What's the probability of any one student's score being 90 or greater?

    a. 12.15%

    b. 17.31%

    c. 23.52%

    d. 15.87%

11. Using the same scenario as in question 10, what's the probability of any one student's score being a failing grade (65 or less)?

    a. 1.68%

    b. 2.42%

    c. 12.40%

    d. 7.14%

12. On your last exam, the class scored an average of 72 with a standard deviation of 12. What is the probability of any one student's z score being between 70 and 80?

    a. 31.61%

    b. 21.42%

    c. 17.67%

    d. 12.14%

13. Which of the following is the most common probability standard used by researchers when conducting analyses?

    a. .01

    b. .05

    c. .10

    d. .50

14. This is defined as "a measure of the lack of symmetry, or 'lopsidedness,' of a distribution."

    a. Skewness

    b. Kurtosis

    c. Mean

    d. Variance

    e. Median

15. This is defined as "a measure that relates to how flat or peaked a distribution appears."

    a. Skewness

    b. Kurtosis

    c. Variance

    d. Mode

    e. Leptokurtic

16. A distribution that is relatively flat compared to a normal distribution is called _____.

    a. skewed

    b. platykurtic

    c. leptokurtic

    d. mesokurtic

EXERCISES

1. Your set of scores has a mean of 5.8 and a standard deviation of 2.3. Calculate the z scores for the following raw scores: 2.1, 5.7, 7.3, and 12.4.

2. You are taking a standardized test, and you want to score in the top 5% of test takers. You know that the mean is 1,000 and the standard deviation is 100. What is the minimum score you need in order to rank in the top 5%?

3. You give your high school calculus students back their most recent test. The mean for the test was 54 with a standard deviation of 11.9. Calculate the raw scores for each of the following z scores: -2.52, .84, and .17.

SHORT-ANSWER/ESSAY QUESTIONS

1. What are some examples of measures in education that you think may be normally distributed? In general, what does the distribution of a measure need to look like for it to be normally distributed?

SPSS QUESTIONS

1. Enter the following data into SPSS: 23, 33, 42, 47, 51, 61, 63, 67, 69, 71. Now, calculate the corresponding z scores. What are your results?

JUST FOR FUN/CHALLENGE YOURSELF

1. First, look up the equation for the normal curve. Now, using this equation, calculate the area under the curve for a z score between 0 and 1.

2. You measure the height of 10 people and come up with an average of 70 inches and a standard deviation of 4 inches. Calculate the t scores for the following cases: 68 inches, 76 inches.

3. If a variable has a mean of 53, a median of 62, and a standard deviation of 4, what is its skewness?

ANSWER KEY

TRUE/FALSE QUESTIONS

1. False. The percentages of scores under the normal curve are constant in the sense that they're the same regardless of the mean and standard deviation of the distribution.

2. True. Because z scores are standard scores, you can compare them across different distributions.

3. True. Values for the area under the normal curve can be represented as probabilities or percentages.

4. False. Because the normal curve is symmetrical, it doesn't matter whether the z score you're looking up is positive or negative. The area under the curve from the mean to a certain z score will be identical for the positive and negative version of that z score.

5. False. A standard score is very different from a standardized score. Standardized scores come from a distribution with a predefined mean and standard deviation, like the distribution of scores on the SAT or GRE.

6. False. This describes a midpoint.

7. False. Given a large enough sample of students you will obtain a normal distribution of scores.

MULTIPLE-CHOICE QUESTIONS

1. (c) 50%

2. (d) 100%

3. (c) 68.26%

4. (d) 95.44%

5. (b) 0

6. (b) positive

7. (b) 4

8. (c) 2.28%

9. (a) 30.85%

10. (d) 15.87%

11. (a) 1.68%

12. (a) 31.61%

13. (b) .05

14. (a) Skewness

15. (b) Kurtosis

16. (b) platykurtic

EXERCISES

1. The four z scores:

$$z = \frac{X - \bar{X}}{s} \Rightarrow \frac{2.1 - 5.8}{2.3} = -1.61$$

$$z = \frac{X - \bar{X}}{s} \Rightarrow \frac{5.7 - 5.8}{2.3} = -0.04$$

$$z = \frac{X - \bar{X}}{s} \Rightarrow \frac{7.3 - 5.8}{2.3} = 0.65$$

$$z = \frac{X - \bar{X}}{s} \Rightarrow \frac{12.4 - 5.8}{2.3} = 2.87$$

2. To achieve a score in the top 5%, you would first need to find the z score such that exactly 5% of scores lies between that z score and the highest possible score (or "infinity"). In other words, 95% of scores are below this score. Therefore, the z score that we are looking for is well above the mean. We know that 50% of scores lie below the mean, so you need to find a z score such that the area between the mean and the z score is 0.45 (this plus 0.50 equals 0.95 or 95%). This corresponds to a z score of approximately 1.645.

   Now calculate:

   $$X = z(s) + \bar{X} \Rightarrow 1.645(100) + 1,000 = 1,164.5$$

3. The three raw scores:

    $-2.52\ (11.9) + 54 = 24$

    $.84\ (11.9) + 54 = 64$

    $.17\ (11.9) + 54 = 56$

SHORT-ANSWER/ESSAY QUESTIONS

1. In this chapter, the examples of IQ and height were presented. Exam scores, as well as final class scores, could also be examples of normally distributed measures. While the mean would probably be around a grade of C, you'd expect to see a smaller number of students get very high grades and a smaller number of students get very low ones. Student weight is another example. You'd expect most students' weight average levels for their age. Smaller percentages of students will be overweight and underweight compared to the average for their age. In essence, the distribution of any normally distributed measure needs to look like the "bell curve": You have a large hump representing typical cases, with a peak equal to the mean. However, you also have smaller numbers of individuals who have more extreme scores, in both the positive and negative tails of the curve.

SPSS QUESTIONS

1. The z scores are presented in the final column in the following screenshot:

	Var1	Zvar1
1	23.00	−1.81871
2	33.00	−1.20635
3	42.00	−.65522
4	47.00	−.34905
5	51.00	−.10410
6	61.00	.50826
7	63.00	.63073
8	67.00	.87567
9	69.00	.99815
10	71.00	1.12062

JUST FOR FUN/CHALLENGE YOURSELF

1.

$$\text{Area from } z_x \text{ to } z_y = \frac{1}{\sqrt{2\pi}} \int_{x}^{y} \Bigl|^{\frac{-z^2}{2}} dz$$

$$= \frac{1}{\sqrt{2\pi}} \int_{0}^{1} \Bigl|^{\frac{-z^2}{2}} dz$$

$$= \frac{1}{\sqrt{2\pi}} (.8556)$$

$$= .3413$$

2. First, we calculate the z scores:

$$z = \frac{X - \bar{X}}{s} \Rightarrow \frac{68 - 70}{4} = -0.5$$

Then,

$$T = z \times 10 + 50 = -0.5 \times 10 + 50 = 45$$

$$T = z \times 10 + 50 = 1.5 \times 10 + 50 = 65$$

3.

$$Sk = \frac{3(\bar{X} - M)}{s} \Rightarrow \frac{3(53 - 62)}{4} = \frac{3(-9)}{4} = -6.75.$$

9 Significantly Significant

What It Means for You and Me

LEARNING OBJECTIVES

- Understand the concept of statistical significance.

- Learn the difference between Type I errors and Type II errors.

- Understand the purpose of inferential statistics.

- Understand the distinction between statistical significance and meaningfulness.

- Learn the eight steps used to apply a statistical test to test any null hypothesis.

- Learn what confidence intervals are.

SUMMARY/KEY POINTS

- Statistical tests are based on probability: You are able to say with a certain level of certainty that there is a difference between groups or a relationship between variables, but you can't say this with 100% absolute certainty.
  - There is the possibility of making an error in judgment.
  - The level of chance or risk that you are willing to take is expressed as a significance level.
  - A significance level of .05 corresponds to a 1 in 20 chance that any differences or relationships found based on statistical tests are not due to the hypothesized reason but are instead due to chance.
  - Researchers should try as much as possible to reduce this likelihood by removing all competing reasons for any differences or relationships. However, error cannot be fully controlled because it is impossible to control for every possible factor.
  - There is always the possibility of error in statistics because the population itself is not directly tested. The sample is tested, and the results are inferred or generalized to the larger population. This inferential process always includes the possibility of error.

- A Type I error occurs when you reject the null hypothesis when there is actually no difference between groups or relationships between variables.
  - The level of statistical significance is equal to the possibility of making a Type I error.
  - Type I errors are represented by the Greek letter alpha, or α.
  - These significance levels are typically set between .01 and .05, with .05 being the most common standard used.
  - A statistical test that is nearly significant can be called "marginally significant." For example, if the probability level is set at .05 and the significance of your result is, say, .052 or .055, it can be reported as a "marginally significant" result.

- A Type II error occurs when you accept a false null hypothesis.
  - This means that you conclude that there is no difference between groups or no relationship between variables when in fact there actually is.
  - Type II errors are represented by the Greek letter beta, or β.
  - Type II errors are not directly controlled for. However, they are related to factors such as sample size; type II errors decrease as the sample size increases.

- Type I and Type II errors cover the scenarios in which errors are made; there are also two scenarios in which you make the correct judgment.
  - You can accept the null hypothesis when the null hypothesis is actually true. This means that you say there is no difference between groups, or no relationship between variables, and you are correct.

○ You can reject the null hypothesis when the null hypothesis is actually false. This means that you say there is a real difference between groups, or relationship between variables, and you are correct.

○ This is also called power, or $1 - \beta$. In other words, power is equal to the value of the Type II error subtracted from 1.

- There is an important difference between statistical significance and meaningfulness. It is possible to have a result that is statistically significant but so small that it is not really meaningful. Determining meaningfulness depends in part on the context of the issue and design of the study.

- While descriptive statistics describe data (by means of tables, charts, etc.), inferential statistics are used to infer something about the population based on a sample's characteristics.

○ The practice of inferential statistics uses a wide variety of statistical tests and analyses to test differences between groups or relationships between variables.

○ Tests of significance are used in inferential statistics. These tests of significance are based on the fact that each null hypothesis can be tested with a particular type of statistical test. Every calculated statistic has a special distribution associated with it. The calculated value is then compared to the distribution to conclude whether the sample characteristics are different from what you would expect by chance.

- Applying a statistical test to any null hypothesis follows eight general steps:
  1. Provide a statement of the null hypothesis.
  2. Set the level of risk associated with the null hypothesis (significance level).
  3. Select the appropriate test statistic.
  4. Compute the test statistic value (also known as the obtained value).
  5. Determine the value (the critical value) needed for rejection of the null hypothesis using the appropriate table of critical values for that particular statistic.
  6. Compare the obtained value with the critical value.
  7. If the obtained value is more extreme than the critical value, the null hypothesis must be rejected.
  8. If the obtained value does not exceed the critical value, the null hypothesis cannot be rejected.

- Confidence intervals represent the best estimate of the range of the population value (or population parameter) based on the sample value (or sample statistic).

○ A higher confidence interval (for example, a 99% confidence interval as compared with a 95% confidence interval) represents a greater degree of confidence, meaning that a wider range of values will be incorporated into the confidence interval.

KEY TERMS

- **Significance level** or **statistical significance**: The level of risk set by the researcher for rejecting a null hypothesis when it is true. In other words, it corresponds to the level of risk that there is actually no relationship between variables when the results of your analysis appear to tell you that there is.

- **Type I error**: The rejection, or the probability of rejection, of a null hypothesis when it is true

- **Type II error**: The acceptance, or the probability of acceptance, of a null hypothesis when it is false

- **Inferential statistics:** A set of tools that are used to infer the results of an analysis based on a sample to the population

- **Test statistic** (aka obtained value): The value that results from the use of a statistical test

- **Critical value:** The value necessary for rejection (or nonacceptance) of the null hypothesis

- **Confidence interval:** The best estimate of the range of a population value given the sample value

TRUE/FALSE QUESTIONS

1. A 99% confidence interval has a larger range than a 95% confidence interval for the same analysis.

2. Using inferential statistics, it is commonly possible to say that you have 100% confidence in a result.

MULTIPLE-CHOICE QUESTIONS

1. A significance level of .05 corresponds to a _____ chance that any differences or relationships found based on a statistical test are not due to the hypothesized reason but are instead due to chance.

   a. 1 in 10

   b. 5 in 10

   c. 1 in 5

   d. 1 in 20

2. Which of the following relates to the rejection of the null hypothesis when it is actually false?

   a. Type I error

   b. Type II error

   c. Power

   d. Statistical significance

3. Which of the following relates to the acceptance of the null hypothesis when it is actually true?

   a. Type I error

   b. Type II error

   c. Power

   d. A correct decision

4. Which of the following relates to the rejection of the null hypothesis when it is actually true?

   a. Type I error

   b. Type II error

    c. Power

    d. Statistical significance

5. Which of the following relates to the acceptance of the null hypothesis when it is actually false?

    a. Type I error

    b. Type II error

    c. Power

    d. Statistical significance

6. Which of the following is calculated by subtracting the Type II error from 1?

    a. Type I error

    b. Type II error

    c. Power

    d. The significance level

7. Type I errors are represented as _____.

    a. α

    b. β

    c. $1 - \beta$

    d. e

8. Type II errors are represented as _____.

    a. α

    b. β

    c. $1 - \beta$

    d. E

9. Your fellow 6th grade teacher believes that this year's class of 6th graders is "just not as smart" as last year's 6th grade class. You conduct a study to see whether there is a significant difference in IQ between the two classes. This year's 6th grade class has an average IQ of 101.7, while last year's 6th grade class has an average IQ of 101.1. You conduct a statistical test, which finds a significant difference between these two groups at the .05 level of significance. In sum, these results are _____.

    a. statistically significant only

    b. meaningful only

    c. statistically significant and meaningful

    d. neither statistically significant nor meaningful

10. When conducting a statistical test, you set the significance level at .05. After running the analysis, you find a significance level of .054. This result is _____.

    a. statistically significant

    b. marginally significant

c. meaningful only

d. none of the above

11. In inferential statistics, you infer from a _____ to a _____.

   a. larger population; smaller sample

   b. smaller sample; larger population

   c. smaller population; larger sample

   d. larger sample; smaller population

12. In terms of z scores, a 95% confidence interval consists of which of the following ranges?

   a. $\pm 1\ z$

   b. $\pm 1.96\ z$

   c. $\pm 2.05\ z$

   d. $\pm 2.56\ z$

13. In terms of z scores, a 99% confidence interval consists of which of the following ranges?

   a. $\pm 1\ z$

   b. $\pm 1.96\ z$

   c. $\pm 2.05\ z$

   d. $\pm 2.56\ z$

14. If a class took an exam and scored a mean of 82 with a standard deviation of 12, what would be the 95% confidence interval?

   a. $12 \pm 2.56(82)$

   b. $82 \pm 2.56(12)$

   c. $12 \pm 1.96(82)$

   d. $82 \pm 1.96(12)$

15. Using the .05 level of significance, which of the following findings is statistically significant?

   a. Lower levels of college attendance have been found among students attending high school in high poverty (low socioeconomic status) areas ($p < .05$).

   b. IQ scores have been found to be increasing steadily since 1900 ($p = .053$).

   c. ACT scores have been steadily increasing in the United States since 1990 ($p = .06$).

   d. No gender differences in scores were found ($p = .12$).

16. Which of the following significance levels gives you the greatest chance of finding a significant result?

   a. .10

   b. .05

   c. .01

   d. .001

17. Which of the following values results from the use of a statistical test?

   a. The critical value

   b. The obtained value

   c. Type I error

   d. Type II error

18. Which of the following is the value necessary for rejection (or nonacceptance) of the null hypothesis?

   a. The critical value

   b. The obtained value

   c. Type I error

   d. Type II error

19. Which of the following is the best estimate of the range of a population value given the sample value?

   a. The critical value

   b. The obtained value

   c. Power

   d. Confidence interval

EXERCISE

1. Come up with a hypothesis in an area of study that interests you. Now, conduct a "mock" statistical test, writing out the eight steps you would use to apply a statistical test to your hypothesis.

SHORT-ANSWER/ESSAY QUESTION

1. What's an example of a result that is statistically significant but not meaningful? Why is it not meaningful?

JUST FOR FUN/CHALLENGE YOURSELF

1. Spend a few minutes reading about statistical power online or in a statistics book. Next, download G*Power, a free software program for power calculations, or find an online calculator that can calculate the power for correlation coefficients. Using a Pearson correlation (bivariate normal model), what is the sample size needed for a two-tailed test if the null correlation (ρ) is 0, the research hypothesis correlation (r) is 0.5, your level of significance is .05, and you want a power of 0.9?

TRUE/FALSE QUESTIONS

1. True. To have a higher level of confidence (i.e., 99% confidence instead of 95% confidence), you would need to incorporate a larger set of values into the range of the confidence interval. This means that a 99% confidence interval must have a larger range than does a 95% confidence interval.

2. False. Inferential statistics uses probability, which in this case means that you can never be completely certain of a result.

MULTIPLE-CHOICE QUESTIONS

1. (d) 1 in 20

2. (c) Power

3. (d) A correct decision

4. (a) Type I error

5. (b) Type II error

6. (c) Power

7. (a) α

8. (b) β

9. (a) statistically significant only

10. (b) marginally significant

11. (b) smaller sample; larger population

12. (b) $\pm 1.96\ z$

13. (d) $\pm 2.56\ z$

14. (d) $82 \pm 1.96(12)$

15. (a) Lower levels of college attendance have been found among students attending high school in high poverty (low socioeconomic status) areas ($p < .05$).

16. (a) .10

17. (b) The obtained value

18. (a) The critical value

19. (d) Confidence interval

EXERCISES

1. As an example, let's say that I'm interested in international math achievement and want to study whether there is a significant difference in math achievement between the United States and Europe. Your answer does not have to be as detailed as mine, and we will get more detailed and specific in the next few chapters. Here are the eight steps that we would use to test this:

1. A statement of the null and research hypotheses:

   The null hypothesis: H_0: $\mu_1 = \mu_2$.

   The research hypothesis: H_1: $\bar{X}_1 \neq \bar{X}_2$.

2. Set the level of risk, or significance, associated with the null hypothesis: 0.05.

3. Select the appropriate test statistic. In this case, we would use the *t*-test for independent means, because we are testing the difference between two separate groups.

4. Compute the test statistic value (obtained value). This goes beyond what we've covered so far, but we will fully complete this step in later chapters, which focus on particular statistical tests. For the sake of argument, say that our obtained value is 3.7.

5. Determine the critical value. Again, to continue with this example, say that our critical value is 2.021.

6. Now, compare the obtained value with the critical value. As you can see, our obtained value is larger than our critical value.

7. Because the obtained value is more extreme than the critical value, we can say that our null hypothesis cannot be accepted.

8. Because the obtained value is larger than the critical value, we must reject the null hypothesis.

SHORT-ANSWER/ESSAY QUESTION

1. A result that is statistically significant but not meaningful could be any difference between groups, or relationship between variables, that has a significance level below .05 (or whatever standard you're using) but is very small. For example, you would have a group difference that is significant but not meaningful if in a psychology class, the class average on the first exam was 87.2, the class average on the second exam was 88.1, and this difference had a significance level below .05. Also, you would have a significant but not meaningful relationship if the correlation between two variables had a significance level below .05 but the strength of the correlation was very weak.

   The reason why these examples do not illustrate a meaningful difference between groups or a meaningful relationship between variables is that the group difference or strength of the relationship is extremely low.

JUST FOR FUN/CHALLENGE YOURSELF

1. The minimum sample size needed in this example is 37.

10 The One-Sample *z*-Test

Only the Lonely

LEARNING OBJECTIVES

- Understand when it is appropriate to use the one-sample z-test.

- Learn how to compute the observed z value for a one-sample z-test.

- Learn how to interpret the z value and understand what it means.

- Be able to work through the eight steps of testing a hypothesis when using the one-sample z-test.

- Learn the basic concepts of effect size.

SUMMARY/KEY POINTS

- A one-sample z-test is used to compare the mean of a sample to the mean of a population.
  - This test is used when only one group is being tested.
  - The obtained value is determined by calculating the one-sample z-test.
  - After the z value is calculated, the critical value is obtained from a table of z scores so that a comparison can be made.
  - Effect size can help you to understand whether a statistically significant result is also meaningful.
    - A small effect size ranges from 0 to .2.
    - A medium effect size ranges from .2 to .5.
    - A large effect size is any value above .5.

KEY TERMS

- **One-sample z-test**: A statistical test used to compare a sample mean to a population mean

- **Standard error of the mean**: An error term that is used as the denominator in the equation for the z value in a one-sample z-test. The standard error of the mean is the standard deviation of all possible means selected from the population.

- **Effect size**: A measure of how different two groups are from one another; it is a measure of the magnitude of a treatment

TRUE/FALSE QUESTIONS

1. A one-sample z-test can be used to compare the mean of two populations.

2. For a one-sample z-test to be significant at the .05 level of significance, you need an obtained z value of at least 1.96 (or below −1.96).

3. If your district assessment specialist wanted to compare the district's average scores on the yearly state standardized test to the average scores across the entire state, she would use a one sample z-test.

MULTIPLE-CHOICE QUESTIONS

1. A one-sample z-test would be used in which one of the following situations?

   a. Comparing two sample means

   b. Comparing two population means

   c. Comparing a sample mean with a population mean

   d. Comparing two sample means with a population mean

2. In a one-sample z-test, _____ groups are being tested.

   a. one

   b. two

   c. three

   d. two or more

3. In the equation for a one-sample z-test, the denominator is known as _____.

   a. the sample mean

   b. the population mean

   c. the standard error of the mean

   d. the standard error of the population

4. Which of the following values represents the standard deviation of all the possible means selected from the population?

   a. The sample mean

   b. The population mean

   c. The standard error of the mean

   d. The standard error of the population

5. You conducted a one-sample z-test, obtaining a value of 4.6 for the z value. What is your conclusion?

   a. I should reject the null hypothesis.

   b. I should accept the null hypothesis.

   c. The result is too close to call.

   d. None of the above is correct.

EXERCISES

1. You are a teacher at a gifted school, and you feel that the newest class of students is even brighter than usual. The mean IQ at your school is 127, and the mean IQ of this new class is 134. In total, there are 32 students in this new class. Also, the standard deviation of the school's IQ is 8. Use the eight steps to test whether this new class of students is significantly more intelligent than the school's student body overall.

2. Interpret the following result: $z = 3.49$, $p < .05$.

SHORT-ANSWER/ESSAY QUESTION

1. Come up with two examples in which a one-sample z-test would be appropriate.

JUST FOR FUN/CHALLENGE YOURSELF

1. A one-sample z-test was conducted in which the sample mean was 76, the population mean was 82, the population standard deviation was 2, and the calculated z value was −15. What was the sample size?

ANSWER KEY

TRUE/FALSE QUESTIONS

1. False. The one-sample z-test is used to compare the mean of a sample with the mean of a population, not to compare the mean of two populations.

2. True. At the .05 level of significance, a two-tailed test has a critical value of 1.96 for the z value.

3. True. A one-sample z-test can be used to compare the district (a sample) to the state (the population) on the state standardized test.

MULTIPLE-CHOICE QUESTIONS

1. (c) Comparing a sample mean with a population mean

2. (a) one

3. (c) the standard error of the mean

4. (c) The standard error of the mean

5. (a) I should reject the null hypothesis.

EXERCISES

1. Here are the eight steps used to test this hypothesis:

    1. The null hypothesis: H_0: $\bar{X} = \mu$

    2. The alternative hypothesis: H_1: $\bar{X} = \mu$

    3. The level of significance: 0.05.

    4. The appropriate test statistic: The one-sample z-test.

    5. Computation of the test statistic value:

$$SEM = \frac{\sigma}{\sqrt{n}} \Rightarrow \frac{8}{\sqrt{32}} = 1.41$$

$$z = \frac{\bar{X} - \mu}{SEM} \Rightarrow \frac{134 - 127}{1.41} = 4.96$$

As this is a one-tailed test, the value needed for rejection of the null hypothesis is ±1.66. This represents the point at which only 5% of scores are higher than this value, corresponding to our significance level of .05. If we had been conducting a two-tailed test, the critical value would have been ±1.96.

6. Comparing the obtained value with the critical value, we see that the obtained value is higher than the critical value.

7. and 8. Because our obtained value is greater than the critical value, we can reject the null hypothesis, which suggests no difference between the sample and population mean. Our results show that this new class is significantly brighter than the school's students overall.

2. First, the z represents the test statistic that was used. The value of 3.49 represents the obtained z value that was calculated as part of conducting the one-sample z-test. Finally, $p < .05$ indicates that we have at least a 95% level of certainty that these two groups (the sample and the population) do differ in regard to their mean values.

SHORT-ANSWER/ESSAY QUESTION

1. These two examples could be any situation in which you are comparing a sample mean to a population mean. For example, a one-sample z-test would be appropriate if you were comparing reading-test scores in one school with scores of the entire nation. Another example could be testing whether the poverty rate in a certain state or neighborhood is significantly different from the national poverty rate.

JUST FOR FUN/CHALLENGE YOURSELF

1. We can solve this using the equation for the one-sample z-test:

$$z = \frac{\bar{X} - \mu}{\frac{\sigma}{\sqrt{n}}} \Rightarrow$$

$$-15 = \frac{76 - 82}{\frac{2}{\sqrt{n}}}$$

$$-15 = \frac{-6}{\frac{2}{\sqrt{n}}}$$

$$-15 = -6\left(\frac{\sqrt{n}}{2}\right)$$

$$-15 = -3\sqrt{n}$$

$$5 = \sqrt{n}$$

$$n = 25$$

11 *t*(ea) for Two

Tests Between the Means of Different Groups

LEARNING OBJECTIVES

- Understand when it is appropriate to use the *t*-test for independent means.

- Learn how to calculate the observed *t* value by hand and with SPSS.

- Learn how to interpret the results of a *t*-test.

- Understand the difference between a significant and meaningful result and how this relates to the effect size.

SUMMARY/KEY POINTS

- Use the *t*-test for independent means when you are looking at the difference between the average scores of two groups on one or more variables and the two groups are independent of one another (i.e., are not related in any way).
  - This test is used when each group is tested only once.
  - There must be only two groups in total.
  - The corresponding test statistic is the *t*-test for independent means.

- After the *t*-test for independent means is conducted, compare the obtained value with the critical value to see whether you have statistical significance.

- There's an important difference between a significant result and a meaningful result. A component of a result's meaningfulness is effect size, which is a measure of the magnitude of the treatment.

- When *t*-tests are used, the measure of effect size is Cohen's *d*.
  - A small effect size ranges from 0 to .20.
  - A medium effect size ranges from .20 to .50.
  - A large effect size is any value above .50.
  - A larger effect size represents a greater difference between the two groups.

KEY TERMS

- **Homogeneity of variance assumption**: An assumption underlying the *t*-test that the amount of variability in both groups is equal

- **Degrees of freedom**: A value that approximates the sample size and is a component of many statistical tests

- **Effect size**: A measure of how different groups are from one another (i.e., a measure of the magnitude of the effect)

- **Pooled standard deviation**: Part of the formula for the effect size that is similar to an average of the standard deviations from both groups

TRUE/FALSE QUESTIONS

1. The observed value $t_{(24)} = 2.35$ is significant at the .05 level (two-tailed test).

2. The *t*-test for independent samples should be used when participants are tested multiple times.

3. The sign of the observed *t* value (i.e., whether it is positive or negative) is always a crucial element in conducting the *t*-test for independent samples.

4. A reading researcher gave 3rd grade students in both charter schools and public schools a standardized test of reading comprehension. This researcher can use a *t*-test for independent samples to analyze her data.

MULTIPLE-CHOICE QUESTIONS

1. Which of the following is significant at the .05 level (two-tailed test)?

    a. $t_{(32)} = 2.01$.

    b. $t_{(212)} = 1.99$.

    c. $t_{(6)} = 2.30$.

    d. $t_{(2)} = 4.10$.

2. Which of the following is significant at the .01 level (one-tailed test)?

    a. $t_{(1)} = 28.90$.

    b. $t_{(16)} = 2.42$.

    c. $t_{(70)} = 2.45$.

    d. $t_{(80)} = 2.37$.

3. The *t*-test for independent samples should be used in which of the following scenarios?

    a. You are comparing more than two groups that are related.

    b. You are comparing exactly two groups that are related.

    c. You are comparing exactly two groups that are unrelated.

    d. You are comparing more than two groups that are unrelated.

4. The assumption that the *t*-test for independent samples makes regarding the amount of variability in each of the two groups is called the _____.

    a. homogeneity of variance assumption

    b. equality of variance assumption

    c. assumption of variance equivalence

    d. variability equality assumption

5. The value used in the equation for the *t*-test for independent means that approximates the sample size is known as the _____.

    a. homogeneity of variances

    b. sample size equivalence

c. degrees of freedom

d. standard deviation

6. Imagine that you look up a critical *t* value in a *t* table and your observed *t* value for a specific degrees of freedom lies between two critical values in the table. To be conservative, you would select the _____.

a. smaller value

b. larger value

c. smallest value in the table

d. largest value in the table

7. After conducting a *t*-test for independent samples, you arrived at the following: $t_{(47)} = 3.41, p < .05$. The degrees of freedom is indicated by which of the following?

a. 47

b. 3.41

c. < .05

d. *p*

e. *t*

8. After conducting a *t*-test for independent samples, you arrived at the following: $t_{(47)} = 3.41, p < .05$. The *t* value is indicated by which of the following?

a. 47

b. 3.41

c. < .05

d. *p*

e. *t*

9. After conducting a *t*-test for independent samples, you arrived at the following: $t_{(47)} = 3.41, p < .05$. The probability is indicated by which of the following?

a. 47

b. 3.41

c. < .05

d. *t*

10. After conducting a *t*-test for independent samples, you arrived at the following: $t_{(47)} = 3.41, p < .05$. The test statistic that was used is indicated by which of the following?

a. 47

b. 3.41

c. < .05

d. *p*

e. *t*

11. An effect size of .25 would be considered _____.

    a. small

    b. medium

    c. large

12. An effect size of .17 would be considered _____.

    a. small

    b. medium

    c. large

13. An effect size of .55 would be considered _____.

    a. small

    b. medium

    c. large

14. If there is no difference between the distributions of scores in two groups, your effect size will be equal to which of the following?

    a. 0

    b. 0.1

    c. 0.5

    d. 1

15. A researcher examined the benefits of language immersion on students' creativity. He gave 40 Spanish immersion 3rd graders and 40 non-immersion 3rd graders a test of creative thinking. He obtained the following results: $t_{(78)} = 4.56$, $p < .05$. Which of the following is the correct interpretation of his results?

    a. There is no difference in creative thinking between students in immersion programs and those not in immersion programs.

    b. There is a significant difference in creative thinking between students in immersion programs and those not in immersion programs.

    c. An independent samples t-test is not the appropriate statistic to use for this study.

    d. Students in immersion programs are more creative in their thinking than those not in immersion programs.

EXERCISES

1. A high school history teacher is interested in exploring whether his afternoon 9th grade class has significantly different test scores than his morning 9th grade class. He has 21 students in each class, and decided to conduct a t-test for independent samples on his last exam. These are his data:

Morning Class Exam Scores	Afternoon Class Exam Scores
87	86
98	73
75	79
88	56
76	72
85	70
92	87
56	59
89	64
85	77
84	74
92	72
87	84
76	96
56	67
76	54
89	88
92	78
78	74
85	81
67	90

Use the eight steps to test the null hypothesis that there is no difference between your two classes.

2. Assuming that the standard deviations of two groups are equal and the standard deviation is 11.2, calculate the effect size if the two groups have means of 81.6 for the morning class and 75.3 for the afternoon class. Is this effect size small, medium, or large?

3. Using the online calculator at www.uccs.edu/lbecker/ calculate the effect size using the data from the previous question. Do your results match?

4. Interpret the following result: $t_{(40)} = 1.81, p > .05$.

5. Explain these results in everyday language as you would if you were explaining them to a person with no statistical experience.

SHORT-ANSWER/ESSAY QUESTIONS

1. Imagine you are reading journal articles for a paper you are writing. The authors of Article One compared two groups on an outcome measure and found a statistically significant difference between the groups. The sample sizes of the groups were small. The authors of

Article Two compared two other groups on a similar outcome measure. The results in this article were also statistically significant. However, the sample sizes in this study were not similar to each other. One group was quite large, and the other group was smaller but still large in comparison to the groups discussed in Article One. The *p* values in the articles were similar, but with such a discrepancy in sample size between the two articles, you are wondering whether the results from both studies are practically meaningful and not just statistically significant. What other result will you look for (or calculate) to answer your question? Explain why you chose that test result and discuss its features.

2. Please refer to page 218 of *Statistics for People . . .* and read the feature titled "So How Do I Interpret . . . ?" You will see that *t* is a negative number. Is the result statistically insignificant because *t* is negative? Please explain your answer. Then describe any other relevant information that could be provided to describe the data used in the analysis.

SPSS QUESTIONS

1. Input the data under the "Exercises" section, question 1, into SPSS. Now, use SPSS to conduct a *t*-test for independent samples. How would you interpret the results? Do your results match those calculated by hand?

2. Open the supplemental data set "Teacher Survey Data" in SPSS. Use SPSS to conduct an independent sample *t*-test comparing the overall satisfaction rates for teachers by degree (1 = bachelor's, 2 = master's). Overall satisfaction rates are the average of the teachers' answers to the question "I am generally satisfied with being a teacher at this school" at two time points. Scores for this item, with your dependent variable in this analysis, range from 1 =strongly disagree, 2 = disagree, 3 = neutral, 4 = agree, to 5 = strongly agree. How would you interpret these results?

JUST FOR FUN/CHALLENGE YOURSELF

1. Generate a list of groups in educational settings that one could use in an independent sample *t*-test to examine group differences (e.g., males vs. females).

2. You study two apple orchards to see whether one produces significantly more fruit than the other. The first orchard produces an average of 14.2 tons of apples per year, while the second produces 17.3 tons per year, on average. The standard deviation of the amount of apples produced per year is 4.2 tons for the first orchard and 2.3 tons for the second orchard. Calculate the effect size using the equation that utilizes the pooled standard deviation.

ANSWER KEY

TRUE/FALSE QUESTIONS

1. True.

2. False. This test should only be used when participants are tested only once.

3. False. The sign of the observed *t* value is not important when the *t*-test for independent samples is nondirectional. The sign of the observed *t* value is important when the *t*-test for independent sample is directional.

4. True. The researcher can use the independent sample *t*-test to compare reading comprehension rates for students at charter schools to reading comprehension rates for students in public schools because she is examining two independent groups for students.

MULTIPLE-CHOICE QUESTIONS

1. (b) $t_{(212)} = 1.99$.

2. (c) $t_{(70)} = 2.45$.

3. (c) You are comparing exactly two groups that are unrelated.

4. (a) homogeneity of variance assumption

5. (c) degrees of freedom

6. (a) smaller value

7. (a) 47

8. (b) 3.41

9. (c) $< .05$

10. (e) *t*

11. (b) medium

12. (a) small

13. (c) large

14. (a) 0

15. (b) There is a significant difference in creative thinking between students in immersion programs and those not in immersion programs.

EXERCISES

1. The eight steps to test this hypothesis would consist of the following:

   1. A statement of the null and research hypotheses:
      The null hypothesis: $H_0: \mu_1 = \mu_1$.
      The research hypothesis: $H_1: \bar{X}_1 \neq \bar{X}_2$.

   2. Set the level of risk associated with the null hypothesis: .05.

   3. Select the appropriate test statistic: the *t*-test for independent means.

4. Compute the test statistic value (obtained value):

$$t = \frac{\bar{X}_1 - \bar{X}_2}{\sqrt{\left[\frac{(n-1)S_1^2 + (n_2 - 1)s_2^2}{n_1 + n_2 - 2}\right]\left[\frac{n_1 + n_2}{n_1 n_2}\right]}} \Rightarrow$$

$$t = \frac{81.57 - 75.29}{\sqrt{\left[\frac{(21-1)11.22^2 + (21-1)11.26^2}{21 + 21 - 2}\right]\left[\frac{21 + 21}{21 \times 21}\right]}}$$

$$= \frac{6.28}{\sqrt{\left[\frac{2517.8 + 2535.8}{44}\right] \times \left[\frac{42}{441}\right]}}$$

$$= 1.81$$

5. Determine the value needed for rejection of the null hypothesis using the appropriate table of critical values. With degrees of freedom of 40, a two-tailed test using the .05 level of significance has a critical value of 2.021.

6. Compare the obtained value with the critical value: In this case, the obtained value is higher than the critical value.

7. and 8. Decision time: Because the obtained value is smaller than the critical value, we choose to accept the null hypothesis that there is no difference between classes. In other words, there is no difference.

2. The effect size is calculated using the following equation:

$$ES = \frac{\bar{X}_1 - \bar{X}_2}{SD} \Rightarrow \frac{81.57 - 75.29}{11.2} = .56$$

The effect size of .56 would be considered large.

3. Using this calculator, we would use 11.2 for the standard deviation of both samples. The calculated effect size (Cohen's d) is identical to what we had calculated by hand, .56.

4. First, t represents the test statistic that was used, 40 is the degrees of freedom, and 1.81 is the obtained value for the test statistic. Finally, $p > .05$ indicates that the probability is greater than 5% that on any one test of the null hypothesis, the two groups do not differ.

5. Exam scores for 9th grade history students show that there is no difference in scores for students who take the class in the morning versus those who take the class in the afternoon.

SHORT-ANSWER/ESSAY QUESTIONS

1. You look for reports of effect size in each of the studies. Effect size measures the values of one group in a study relative to those of another. This indicates the amount of overlap between the values of each group. A large effect size indicates a relative lack of overlap between the two groups, which, in turn, demonstrates the practical meaningfulness of a result. Even if the authors of Article One and Article Two did not report effect sizes, in the case of t-tests, effect size is easy to calculate. You could do the math to see whether the studies have meaningful results, despite the large differences in sample sizes.

2. No, it is not the negative value that makes this result nonsignificant. The t value is small, and the p value is not less than .05. As with a correlation coefficient, significance is not determined by the positive or negative value of a t value but rather by whether the absolute value of t is greater than the critical value. You could also provide the mean and standard deviation of each group, which would help people to understand the data better.

SPSS QUESTIONS

1. The following is the output from SPSS for this *t*-test for independent samples:

Group Statistics					
Class		**N**	**Mean**	**Std. Deviation**	**Std. Error Mean**
Exam	Morning	21	81.5714	11.21861 11.26118	2.44810 2.45739
	Afternoon	21	75.2857		

Independent Samples Test								
		Levene's Test for Equality of Variances		*t*-test for Equality of Means				
		F	Sig.	*t*	df	Sig. (2-tailed)	Mean Difference	Std. Error Difference
Exam	Equal variances assumed	.001	.979	1.812	40	.077	6.28571	3.46871
	Equal variances not assumed			1.812	39.999	.077	6.28571	3.4671

These results match those that were calculated by hand. The *t* value was not found to be statistically significant at the .05 level, meaning that the null hypothesis—which states that there is no difference between the two groups of respondents—should be accepted.

2. The following is the output from SPSS for this *t*-test for independent samples:

Group Statistics					
	Degree	**N**	**Mean**	**Std. Deviation**	**Std. Error Mean**
Overall Satisfaction	Bachelor's	27	3.8148	.87868	.16910
	Master's	43	4.1977	.69983	.10672

Independent Samples Test										
		Levene's Test for Equality of Variances		*t*-test for Equality of Means					95% Confidence Interval of the Difference	
		F	Sig.	*t*	*df*	Sig. (2-tailed)	Mean Difference	Std. Error Difference	Lower	Upper
Overall Satisfaction	Equal variances assumed	1.124	.293	−2.017	68	.048	−.38286		−.76167	−.00405
	Equal variances not assumed			−1.915	46.290	.062	−.38286	.19996	−.78530	.01958

The Levene's test for this analysis reveals that the assumption of equal variance is not violated, as the t value was found to be statistically significant at the .05 level, meaning that the null hypothesis should be rejected. For this study we can conclude that there is a difference in overall satisfaction levels for teachers based on degree. Looking at the means we can conclude that teachers with a master's degree have higher overall satisfaction rates (M54.197) than those with a bachelor's degree (M53.81)

JUST FOR FUN/CHALLENGE YOURSELF

1. Here are a few examples (in reality this list could be endless):

Group 1	Group 2
Students who learned content through a fieldtrip	Students who learned content in the classroom
Community college	4-year university
Students with an IEP	Students without an IEP
Teachers with master's degrees	Teachers without master's degrees

2. The effect size using the pooled standard deviation is calculated using the following equation:

$$ES = \frac{\overline{X}_1 \overline{X}_2}{\sqrt{\dfrac{\sigma_1^2 + \sigma_2^2}{2}}} \Rightarrow \frac{14.2 - 17.3}{\sqrt{\dfrac{4.2^2 + 2.3^2}{2}}} = -.92$$

12 *t*(ea) for Two (Again)

Tests Between the Means of Related Groups

LEARNING OBJECTIVES

- Understand the purpose of the *t*-test for dependent means and when it should be used.
- Learn how to compute the observed *t* value for this test by hand and by using SPSS.
- Learn how to interpret the *t* value and understand what it means.

SUMMARY/KEY POINTS

- The *t*-test for dependent means is used to determine whether there is a significant difference in the scores of a group of respondents who are tested at two points in time.
- The *t*-test for dependent means can also be used to determine whether there is a significant difference in the scores of two groups of respondents whom the researcher intentionally matched on characteristics important to the study. This study design is called matched pairs. This *t*-test is also known as the *t*-test for paired samples or the *t*-test for correlated samples.
- When *t*-tests are used, the measure of effect size is Cohen's *d*:
  - A small effect size ranges from 0 to .20.
  - A medium effect size ranges from .20 to .50.
  - A large effect size is any value above .50.
  - A larger effect size represents a greater difference between the two groups.

KEY TERMS

- **_t_-test for dependent means:** A test of the difference between means in a group of respondents who were tested at two points in time. Also known as paired or correlated samples *t*-tests.

TRUE/FALSE QUESTIONS

1. The following observed value: $t(37) = 5.62$ is significant at the .05 level (two-tailed test).

2. The *t*-test for dependent means should be used when participants are tested two or more times.

3. The effect size in a *t*-test for independent means is different from the effect size in a *t*-test for dependent means.

4. A researcher examined a group of students in 7th grade and different group of students in 9th grade. This research should compare the two groups using a *t*-test for dependent samples.

MULTIPLE-CHOICE QUESTIONS

1. The *t*-test for dependent means should be used in which of the following scenarios?

   a. You are comparing more than two groups that are related.

   b. You are comparing exactly two groups that are related.

   c. You are comparing exactly two groups that are unrelated.

   d. You are comparing more than two groups that are unrelated.

2. If you are running a *t*-test for dependent means on a group of 25 students, your degrees of freedom will be which of the following?

   a. 25

   b. 26

   c. 24

   d. 12.5

3. If you are hypothesizing that students' posttest scores will be lower than their pretest scores, you should use a _____.

   a. one-tailed test

   b. two-tailed test

   c. *t*-test for independent means

   d. descriptive statistics

4. After conducting a *t*-test for dependent means, you arrived at the following: $t_{(22)} = 12.41$, $p < .05$. The degrees of freedom is indicated by which of the following?

   a. *t*

   b. $p < .05$

   c. 22

   d. 12.41

5. After conducting a *t*-test for dependent means, you arrived at the following: $t_{(22)} = 12.41$, $p < .05$. The test statistic is indicated by which of the following?

   a. *t*

   b. $p < .05$

   c. 22

   d. 12.41

6. After conducting a *t*-test for dependent means, you arrived at the following: $t_{(22)} = 12.41$, $p < .05$. The obtained value is indicated by which of the following?

   a. *t*

   b. $p < .05$

   c. 22

   d. 12.41

7. After conducting a *t*-test for dependent means, you arrived at the following: $t_{(22)} = 12.41$, $p < .05$. The probability is indicated by which of the following?

   a. *t*

   b. $p < .05$

   c. 22

   d. 12.41

8. An effect size of .63 would be considered _____ and _____?

   a. significant; meaningful

   b. not significant; meaningful

   c. not significant; not meaningful

   d. significant; not meaningful

EXERCISE

1. Using the following data, conduct the eight steps of hypothesis testing to see whether there is a difference between these individuals' incomes before and after they return to school to get a master's degree.

Before ($1,000)	After ($1,000)
14	22
24	24
35	44
57	59
35	30
34	67
88	95
57	65

SHORT-ANSWER/ESSAY QUESTION

1. Your school is attempting a new method to deter tardiness. The school collects data on the number of tardy students for one month then implements the new punishment for tardiness and collects data on the number of tardy students for another month. After conducting a *t*-test for dependent means, you found the following result: $t_{(12)} = 0.12$, $p > .05$. Interpret this result.

SPSS QUESTIONS

1. Input the data in the "Exercises" section into SPSS and run a *t*-test for dependent means. How would you interpret the results? Do your results match those calculated by hand?

2. Open the supplemental data set "Teacher Survey Data" in SPSS. In this data set you will find the variables "SurveyItemA10" and "SurveyItemB10." These items represent teachers' responses on a 1–5 scale ranging from 1 = strongly disagree, 2 = disagree, 3 = neutral, 4 = agree, to 5 = strongly agree to the item "i am very stressed at work" at two time points. The first time point (SurveyItemA10) was in the middle of the school year and the second (SurveyItemB10) was during the last week of school. Use SPSS to conduct a dependent sample *t*-test comparing stress for teachers based on time of the year. How would you interpret these results?

ANSWER KEY

TRUE/FALSE QUESTIONS

1. True.

2. False. This test should only be used when participants are tested exactly twice.

3. False. Effect size is measured in the same way for both independent and dependent means *t*-tests. The formula to determine effect size is also the same for both types of *t*-tests.

4. False. These two groups would consist of different people and therefore a *t*-test for independent samples should be performed.

MULTIPLE-CHOICE QUESTIONS

1. (b) You are comparing exactly two groups that are related.

2. (c) 24

3. (a) one-tailed test

4. (c) 22

5. (a) *t*

6. (d) 12.41

7. (b) $p < .05$

8. (a) significant; meaningful

EXERCISE

1. The eight steps to test this hypothesis would consist of the following:

   1. A statement of the null and research hypotheses:

      The null hypothesis: H_0: $\mu_{pretest}$ = $\mu_{posttest}$.
      The research hypothesis: H_1: $\bar{X}_{pretest}$ ≠ $\bar{X}_{posttest}$.

   2. Set the level of risk associated with the null hypothesis: .05.

   3. Select the appropriate test statistic: the t-test for dependent means.

   4. Compute the test statistic value (obtained value):

   $$t = \frac{\sum D}{\sqrt{\dfrac{n\sum D^2(\sum D)^2}{n-1}}} \Rightarrow \frac{62}{\sqrt{\dfrac{8\times 1,376 - 3,844}{7}}} = 1.938.$$

   5. Determine the value needed for rejection of the null hypothesis using the appropriate table of critical values. With degrees of freedom of 7, a two-tailed test using the .05 level of significance has a critical value of 2.365.

   6. Compare the obtained value with the critical value: In this case, the obtained value is lower than the critical value.

   7. and 8. Decision time: Because the obtained value is lower than the critical value, we are not able to reject the null hypothesis that there is no difference between "before" and "after" values for salary. In other words, the null hypothesis is the most attractive explanation—it holds.

SHORT-ANSWER/ESSAY QUESTION

1. First, you can see that the test statistic used here was the t-test for dependent means. The degrees of freedom was 12, and the obtained t value was found to be 0.12. Finally, this result had a probability level above .05, meaning that no significant differences were found between the previous punishment for tardiness and the new punishment for tardiness.

SPSS QUESTIONS

1. The following consists of the SPSS output for this test:

Paired Samples Statistics		Mean	N	Std. Deviation	Std. Error Mean
Pair 1	Before	43.0000	8	23.38498	8.26784
	After	50.7500	8	25.38700	8.97566

Paired Samples Correlations		N	Correlation	Sig.
Pair 1	Before & After	8	.896	.003

Paired Samples Test								
	Paired Differences							
				95% Confidence Interval of the Difference				Sig. (2-tailed)
	Mean	Std. Deviation	Std. Error Mean	Lower	Upper	*t*	*df*	
Pair 1 Before - After	−7.75000	11.31055	3.99888	217.20586	1.70586	21.938	7	.094

In this test, the two-tailed significance level is found to be .094; because this value is not below .05, we're not able to reject the null hypothesis that there is no difference between "before" and "after" scores. The calculated result for the *t* value is identical to the result calculated by hand.

2. The following consists of the SPSS output for this test:

Paired Samples Statistics				
	Mean	N	Std. Deviation	Std. Error Mean
Pair 1 I am very stressed at work (midyear)	3.6000	70	.93870	.11220
I am very stressed at work (end of year)	2.5429	70	.79282	.09476

Paired Samples Correlations			
	N	Correlation	Sig.
Pair 1 I am very stressed at work (midyear) & I am very stressed at work (end of year)	70	.471	.000

Paired Samples Test								
	Paired Differences							
				95% Confidence Interval of the Difference				Sig. (2-tailed)
	Mean	Std. Deviation	Std. Error Mean	Lower	Upper	*t*	*df*	
Pair 1 I am very stressed at work (midyear) - I am very stressed at work (end of year)	1.05714	.89904	.10746	.84277	1.27151	9.838	69	.000

The *t* value was found to be statistically significant at the .05 level, meaning that the null hypothesis should be rejected. For this study we can conclude that there is a difference in overall stress for teachers based on time. Looking at the means we can conclude that teachers have higher stress levels (M = 3.60) in the middle of the year than they do the last week of school (M = 2.54).

13 Two Groups Too Many?

Try Analysis of Variance

LEARNING OBJECTIVES

- Understand what an analysis of variance is and when it should be used.

- Understand the difference between the *t*-test and ANOVA.

- Learn how to compute and interpret the *F* statistic, both by hand and by using SPSS.

- Learn how to compute η^2 (eta squared), the effect size for ANOVA.

SUMMARY/KEY POINTS

- Analysis of variance is used to test whether there is a significant difference in the mean of some dependent variable on the basis of group membership. Unlike the *t*-test, ANOVA can be used to test the difference between two groups as well as more than two groups of respondents. When the *t*-test is calculated between multiple pairings of the same data, the Type I error rate goes up; therefore, the ANOVA should be conducted instead.
  - The corresponding test statistic for ANOVA is the *F*-test.
  - The type of ANOVA covered in this chapter (the simple analysis of variance) is used for participants who are tested only once.

- In essence, in an ANOVA, the variance due to differences in scores is separated into variance that's due to differences between individuals within groups and variance due to differences between groups. Then, these two types of variance are compared.

- There are several types of ANOVA.
  - The simple analysis of variance, or one-way analysis of variance, includes only one factor or treatment variable in the analysis. Following are guidelines for when the one-way ANOVA is the correct statistic to use:
    - There is only one dimension or treatment.
    - There are more than two levels of the grouping factor.
    - One is looking at differences across groups in average scores.
  - A factorial design includes more than one treatment factor. An example would be an outcome measure of a language-development test score, with the factors being number of hours of preschool participation (three possibilities) and placement in Head Start (yes or no).

- ANOVA is an omnibus test, meaning that it tests overall differences between groups and does not tell you which groups are higher or lower than others.
  - If you need to use ANOVA instead of a *t*-test when comparing only two groups, an *F* value for two groups is equal to a *t* value squared, or $F = t^2$.
  - Post hoc comparisons can be used to determine whether there are significant differences between specific groups. The Bonferroni is an example of a frequently used post hoc test.
  - For ANOVA, the measure of effect size is η^2 (eta squared). The scale of how large the effect size is as follows:
    - A small effect size is about .01.
    - A medium effect size is about .06.
    - A large effect size is about .14.

KEY TERMS

- **Analysis of variance**: A test for the difference between two or more means

- **Simple analysis of variance** (aka one-way analysis of variance): A type of ANOVA in which one factor or treatment variable (such as group membership) is being explored

- **Factorial design**: A more complex type of ANOVA in which more than one treatment factor (such as group membership and gender) is being explored

- **Post hoc comparisons**: In relation to ANOVA, tests that are done in addition to ANOVA in order to look at specific group comparisons

- η^2 **(eta squared)**: The measure of effect size used for ANOVA F-tests

TRUE/FALSE QUESTIONS

1. The F statistic obtained from running an analysis of variance will tell you which groups have significantly higher, or significantly lower, scores as compared with every other group.

2. All F-tests are nondirectional.

3. A researcher is studying the school participation levels of parents with children in elementary school, middle school, and high school. This researcher should use a one-way ANOVA to examine participation levels.

MULTIPLE-CHOICE QUESTIONS

1. ANOVA is appropriate for which of the following situations?

   a. Two groups of teachers are tested only once.

   b. Two groups of parents are tested twice.

   c. Three groups of students are tested only once.

   d. Four groups of administrators are tested twice.

   e. Both a and c are correct.

2. The corresponding test statistic for the ANOVA is the _____.

   a. p statistic

   b. t statistic

   c. F statistic

   d. r statistic

3. Which of the following is the type of ANOVA used when there is only one treatment factor?

   a. Simple analysis of variance

   b. Factorial design

   c. Post hoc comparisons

   d. Independent-samples t-test

4. Which of the following is the type of ANOVA used when there are two or more treatment factors?

   a. Simple analysis of variance

   b. Factorial design

   c. Post hoc comparisons

   d. Independent-samples *t*-test

5. Which of the following is the type of test used to look at specific group comparisons?

   a. Simple analysis of variance

   b. Factorial design

   c. Post hoc comparisons

   d. Independent-samples *t*-test

6. If you ran a factorial ANOVA using gender and social class, with the latter categorized as low, medium, or high, your factorial design would be which of the following?

   a. 2×2

   b. 3×2

   c. 1×1

   d. 1×3

7. An effect size of .12 would be considered _____.

   a. small

   b. medium

   c. large

8. An effect size of .03 would be considered _____.

   a. small

   b. medium

   c. large

9. An effect size of .16 would be considered _____.

   a. small

   b. medium

   c. large

10. If there is no difference between the distributions of scores in the groups compared within the ANOVA, your effect size will be equal to which of the following?

    a. 0

    b. 0.1

    c. 0.5

    d. 1

EXERCISES

1. If you have *MS* between groups of 2.4 and *MS* within groups of 0.3, what would your calculated *F* statistic be?

2. You are interested in testing whether three groups of respondents—office workers, students, and rock 'n' roll musicians—significantly differ in their self-rated happiness. The happiness score is calculated on a scale of zero to 100, with 100 indicating the highest possible happiness. Using the following data, conduct the eight steps of hypothesis testing.

Office Workers	Students	Rock 'n' Roll Musicians
23	47	88
43	77	98
56	84	78
89	55	76
45	67	82
55	76	95
23	45	79
33	67	85
27	87	94
26	66	87

3. Using the results from the previous question, construct an *F* table (an example is presented on page 252 of the text).

SHORT-ANSWER/ESSAY QUESTIONS

1. What is the critical *F* statistic value if your total sample size is 50 and you are comparing teachers, paraprofessional, and non-teaching staff (at the .05 level of significance)?

2. What is the critical *F* statistic value at the .05 level of significance for the following: $F(4, 70)$?

3. How would you interpret $F(2, 30) = 32.60$, $p < .05$?

SPSS QUESTION

1. Run an analysis of variance using the data in the "Exercises" section, question 2 (additionally, run the Bonferroni post hoc comparison). Do your results match those calculated by hand? How would you interpret these results?

2. Open the supplemental data set "Teacher Survey Data" in SPSS. Use SPSS to conduct a one-way ANOVA on the overall satisfaction of teachers based on the number of locations in which they teach (1 = 1 location, 2 = 2 locations, 3 = 3 locations). Also, run a Bonferroni comparison and get descriptive statistics. How would you interpret these results?

JUST FOR FUN/CHALLENGE YOURSELF

1. If you are only comparing two groups of respondents, and the t value is found to be 2.30, what will the F statistic be?

2. If someone performs multiple t-tests, and the initial Type I error rate is .05 and 8 comparisons are made, what will the actual Type I error rate be?

ANSWER KEY

TRUE/FALSE QUESTIONS

1. False. Post hoc comparisons are necessary for this—the F statistic is nondirectional.

2. True.

3. True. Parents would be put into one of three groups (elementary, middle school, or high school parents) and school participation levels (the dependent variable) can then be compared across the three groups of parents using a one-way ANOVA with three groups.

MULTIPLE-CHOICE QUESTIONS

1. (c) Three groups of students are tested only once.

2. (c) F statistic

3. (a) Simple analysis of variance

4. (b) Factorial design

5. (c) Post hoc comparisons

6. (b) 3×2

7. (b) medium

8. (a) small

9. (c) large

10. (a) 0

EXERCISES

1. This would be calculated in the following way:

$$F = \frac{MS_{between}}{MS_{within}} = \frac{2.4}{0.3} = 8.0$$

2. The eight steps of hypothesis testing:

   1. State the null and research hypotheses:

   $H_0: \mu_1 = \mu_2 = \mu_3$
   $H_1: \bar{X}_1 \neq \bar{X}_2 \neq \bar{X}_3$

2. Set the level of significance: 0.05.

3. Select the appropriate test statistic: a simple ANOVA.

4. Compute the test statistic value:

	Office Workers	Students	Musicians
n	10	10	10
$\sum X$	420	671	862
$\sum(X^2)$	21,508	46,923	74,828
$(\sum X)^2/n$	17,640.0	45,024.1	74,304.4

$\sum\sum X = 1,953.$

$(\sum\sum X)^2/N = 127,140.3.$

$\sum\sum(X^2) = 143,259.$

$\sum(\sum X)^2/n = 139,968.5.$

$SS_{Between} = \sum(\sum X)^2/n - (\sum\sum X)^2/N = 139,968.5 - 127,140.3 = 9.828.2.$

$SS_{Within} = \sum\sum(X^2) - \sum(\sum X)^2/n = 143,259 - 139,968.5 = 6,290.5.$

$MS_{Between} = SS_{Between}/(k-1) = 9.828.2/2 = 4914.1.$

$MS_{Within} = SS_{Within}/(N-k) = 6,290.5/(30 - 3) = 232.98.$

$F = MS_{Between}/MS_{Within} = 4914.1/232.98 = 21.09223.$

5. Determine the value needed to reject the null hypothesis: Critical $F_{(2, 27)} = 3.36$.

6. Compare the obtained value with the critical value: The obtained value, 21.09, is larger than the critical value of 3.36.

7. and 8. Decision time: Because the obtained value is greater than the critical value, we would reject the null hypothesis that states there is no difference between these groups.

3. The F table:

Source	Sum of Squares	df	Mean Sum of Squares	F
Between groups	9,828.2	2	4,914.1	21.09
Within groups	6,290.5	27	232.98	
Total	16,118.7	29		

SHORT-ANSWER/ESSAY QUESTIONS

1. The critical F statistic in this case would be 3.21. This result is obtained by being more conservative and looking at the value that corresponds to a degrees of freedom of 45.

2. The critical F statistic in this case is 2.51.

3. First, the F represents the test statistic that was used. Then, 2 and 30 represent the degrees of freedom for the between-group and within-group estimates, respectively. The value of 32.60 represents the obtained value, which was arrived at by using the formula for the F statistic. Finally, $p < .05$ indicates that the probability is less than 5% that the average

scores between groups differ due to chance as opposed to the effect of the treatment or group membership. This also indicates that the research hypothesis should be preferred over the null hypothesis, as there is a significant difference.

SPSS QUESTION

1. The SPSS output is shown here:

One-Way

ANOVA

Source	Sum of Squares	df	Mean Squares	F	Sig.
Between groups	9,828.200	2	4,914.100	21.092	.000
Within groups	6,290.500	27	232.981		
Total	16,118.700	29			

Post Hoc Tests

MULTIPLE COMPARISONS

Score

Bonferroni

(I) Group	(J) Group		Mean Difference (I – J)	Std. Error	Sig.	95% Confidence Interval	
						Lower Bound	Upper Bound
dimension2	1 dimension3	2	−25.100*	6.826	.003	−42.52	−7.68
		3	−44.200*	6.826	.000	−61.62	−26.78
	2 dimension3	1	25.100*	6.826	.003	7.68	42.52
		3	−19.100*	6.826	.028	−36.52	−1.68
	3 dimension3	1	44.200*	6.826	.000	26.78	61.62
		2	19.100*	6.826	.028	1.68	36.52

* The mean difference is significant at the 0.05 level.

These results, in regard to the one-way ANOVA, do match those calculated by hand. First, the *F*-test was found to be statistically significant, indicating that the null hypothesis should be rejected in favor of the research hypothesis. In other words, there is a significant difference between groups. Additionally, the Bonferroni post hoc comparison finds a significant difference among all three groups of respondents. Focusing on the final row, which compares group 3 (musicians) against groups 1 (office workers) and 2 (students), and looking at the mean difference and significance columns, we can see that musicians are significantly happier than both students and office workers (no surprise). Additionally, the only other comparison, which was between groups 1 and 2, is shown in the first row of results. Here, we can see that office workers are significantly less happy than are students (again, no surprise here).

2. The SPSS output is shown here:

DESCRIPTIVES
Overall Satisfaction

	N	Mean	Std. Deviation	Std. Error	95% Confidence Interval for Mean		Minimum	Maximum
					Lower Bound	Upper Bound		
1.00	57	4.1579	.69538	.09211	3.9734	4.3424	2.00	5.00
2.00	5	3.7000	1.15109	.51478	2.2707	5.1293	2.00	4.50
3.00	8	3.5000	1.00000	.35355	2.6640	4.3360	1.50	4.50
Total	70	4.0500	.79011	.09444	3.8616	4.2384	1.50	5.00

ANOVA
Overall Satisfaction

	Sum of Squares	df	Mean Square	F	Sig.
Between Groups	3.696	2	1.848	3.144	.050
Within Groups	39.379	67	.588		
Total	43.075	69			

MULTIPLE COMPARISONS
Dependent Variable: Overall Satisfaction
Bonferroni

(I) Number of Teaching Locations	(J) Number of Teaching Locations	Mean Difference (I – J)	Std. Error	Sig.	95% Confidence Interval	
					Lower Bound	Upper Bound
1.00	2.00	.45789	.35758	.614	−.4202	1.3359
	3.00	.65789	.28945	.079	−.0529	1.3687
2.00	1.00	−.45789	.35758	.614	−1.3359	.4202
	3.00	.20000	.43706	1.000	−.8732	1.2732
3.00	1.00	−.65789	.28945	.079	−1.3687	.0529
	2.00	−.20000	.43706	1.000	−1.2732	.8732

The *F*-test was found not to be statistically significant, with a significance level of .05 and not the required less than .05. This indicates that the null hypothesis should be accepted. In other words, there is no significant difference between teacher satisfaction based on the number of locations in which they teach. The Bonferroni post hoc comparison also shows no significant difference in satisfaction for teachers based on number of teaching locations.

JUST FOR FUN/CHALLENGE YOURSELF

1. To obtain the *F* statistic in this case, we simply need to square the *t* value: This gives us a calculated *F* statistic of 5.29.

2. If someone performs multiple *t*-tests, and the initial Type I error rate is .05 and 8 comparisons are made, what will the actual Type I error rate be? This can be calculated by using the following equation:

$$\text{True Type I error} = 1 - (1 - \alpha)^k = 1 - (1 - .05)^8 = 0.34$$

14 Two Too Many Factors

Factorial Analysis of Variance—A Brief Introduction

LEARNING OBJECTIVES

- Learn when it is appropriate to use the factorial analysis of variance.

- Understand the distinction between main effects and interaction effects and what they indicate.

- Learn how to use SPSS to conduct a factorial analysis of variance.

- Learn how to compute the effect size for factorial analysis of variance.

SUMMARY/KEY POINTS

- The factorial analysis of variance, or two-way analysis of variance, is used when you have more than one factor. Factors are also referred to as independent or treatment variables.
  - This type of ANOVA can test the significance of the main effects of each independent variable, as well as the significance of the interaction effect between independent variables.
  - This type of ANOVA is used when participants are tested only once.
  - The test statistic used is the factorial analysis of variance.
  - There will be at least three hypotheses: a main effect for each factor and an interaction.
  - Factorial ANOVA allows for more complex research questions to be asked.
  - For factorial ANOVA, effect size is measured by ω^2 (omega squared), and the point is still to make a judgment about the magnitude of an observed difference.

KEY TERMS

- **Factorial analysis of variance** (aka two-way analysis of variance): The type of ANOVA that is used when there is more than one independent variable or factor

- **Main effect**: In analysis of variance, a significant effect of a factor, or independent variable, on the outcome variable

- **Source table**: A listing of sources of variance in an analysis of variance summary table. This will be produced in SPSS output and will show the obtained F value and level of significance for each factor and the interaction, as well as other information.

- **Interaction effect**: The varying effect of one independent variable on the dependent variable depending on the level of a second independent variable. For example, a significant interaction between exercise treatment (low- or high-impact exercise) and gender indicates that females lose more weight than males in the high-impact treatment condition and males lose more weight than females under the low-impact condition.

- ω^2 **(omega squared)**: The name of the effect size used for factorial ANOVA

TRUE/FALSE QUESTIONS

1. A factorial analysis of variance can be used only in the case where you have two independent, or treatment, variables.

2. Interaction effects are always significant when all main effects are significant.

3. If the authors of an article you read do not provide the value of ω^2 but do provide a full source table/*F* table, you can calculate the effect size yourself.

MULTIPLE-CHOICE QUESTIONS

1. A factorial analysis of variance should be used in which of the following situations?

    a. You have two independent variables, and participants are tested more than once.

    b. You have one independent variable, and participants are tested more than once.

    c. You have two independent variables, and participants are tested only once.

    d. You have three independent variables, and participants are tested more than once.

2. Which of the following statements is true about a factorial analysis of variance?

    a. The number of main effects will be equal to the number of independent variables.

    b. The number of main effects will be equal to the number of independent variables plus the number of interaction effects.

    c. The number of main effects will be equal to the number of interaction effects.

    d. The number of main effects will be equal to the number of independent variables minus the number of interaction effects.

3. What does the following constitute: "The effect of being upper-class, middle-class, or lower-class is different for males and females"?

    a. An interaction effect

    b. A main effect

    c. Either an interaction effect or a main effect

    d. None of the above

4. When plotted on a graph, a significant interaction effect is indicated by _____.

    a. parallel lines

    b. lines that cross/are not parallel

    c. either a or b

    d. none of the above

EXERCISE

1. Draw an example of a graph that illustrates a strong interaction effect. Now, draw a graph that illustrates no interaction effect at all.

SHORT-ANSWER/ESSAY QUESTION

1. In what situations would a factorial ANOVA be preferred over the one-way ANOVA and why?

SPSS QUESTIONS

1. Using the following data, use the eight steps of hypothesis testing to determine the main effects of group membership and the interaction between independent variables. Additionally, write up the three results as they would be printed in a journal article or report (see page 272 of the text for an example). The dependent variable is health scores, while the independent variables are gender and social class. With regard to health scores, higher values indicate better overall physical health.

Note: Include both independent variables as fixed factors.

Health Score	Gender	Social Class
98	Female	Upper
88	Male	Upper
87	Female	Upper
85	Female	Middle
75	Female	Upper
74	Male	Upper
72	Male	Middle
71	Female	Middle
65	Male	Middle
55	Male	Upper
47	Male	Middle
33	Male	Lower
22	Female	Lower
10	Male	Upper
5	Male	Lower

2. Open the supplemental data set "Teacher Survey Data" in SPSS. Use SPSS to conduct a two-way ANOVA to determine the main effects of "working another job" and "degree" on the dependent variable of overall stress in teachers. Be sure to request descriptive statistics in the "options" section of your univariate analysis and conduct a Bonferroni posthoc test on the variable "working another job." Then discuss what these results mean.

JUST FOR FUN/CHALLENGE YOURSELF

1. Using the following data, conduct a multivariate analysis of variance in SPSS and interpret the results. The two dependent variables are health scores and lifestyle attitudes. Higher values on the health score variable indicate better general physical health, while higher scores on the lifestyle attitudes variable indicate more positive and more healthy attitudes.

Lifestyle Attitudes	Health Score	Gender	Social Class
88	98	Female	Upper
76	88	Male	Upper
98	87	Female	Upper
76	85	Female	Middle
67	75	Female	Upper
86	74	Male	Upper
65	72	Male	Middle
67	71	Female	Middle
65	65	Male	Middle
43	55	Male	Upper
54	47	Male	Middle
22	33	Male	Lower
32	22	Female	Lower
11	10	Male	Upper
2	5	Male	Lower

ANSWER KEY

TRUE/FALSE QUESTIONS

1. False. The factorial analysis of variance can be used when you have two or more than two independent/treatment variables.

2. False. Whether the interaction effects are significant will not depend on the significance of the main effects.

3. True. To calculate ω^2, you need $SS_{between}$, $df_{between}$, MS_{within}, and SS_{total}, all of which can be found in the source table of a factorial ANOVA F-test.

MULTIPLE-CHOICE QUESTIONS

1. (c) You have two independent variables, and participants are tested only once.

2. (a) The number of main effects will be equal to the number of independent variables.

3. (a) An interaction effect

4. (b) lines that cross/are not parallel

EXERCISE

1. This first graph presents an example of a strong interaction effect; the lines cross and have very different slopes.

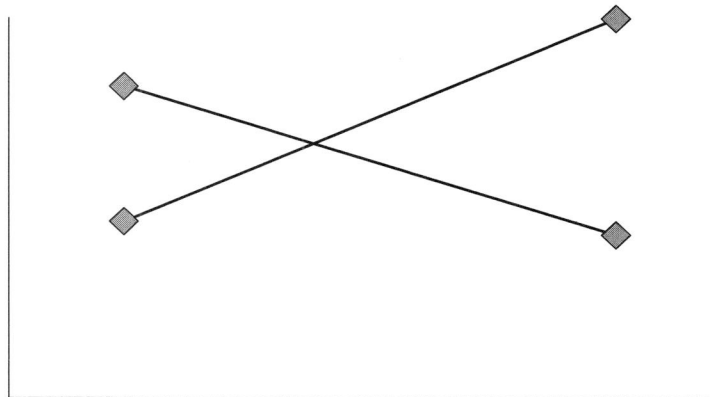

This next graph illustrates no interaction effect, as the lines are parallel to each other.

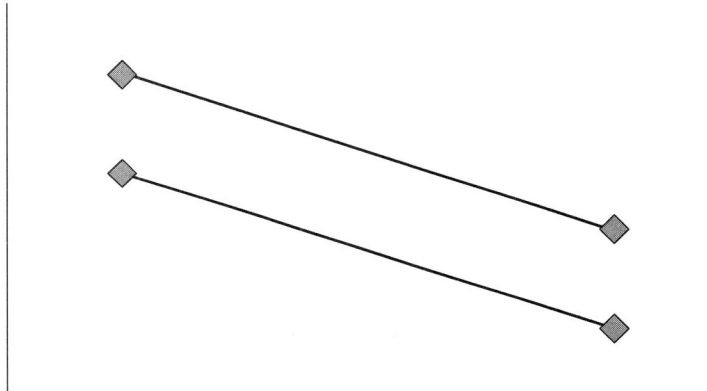

SHORT-ANSWER/ESSAY QUESTION

1. A factorial ANOVA is preferred over a one-way ANOVA whenever there is more than one independent, or treatment, variable. The factorial ANOVA is preferred because a one-way ANOVA can only incorporate a single independent variable into the analysis.

SPSS QUESTIONS

1. The eight steps of hypothesis testing:

    1. State the null and research hypotheses:

        The null hypotheses:

        For gender: H_0: $\mu_{male} = \mu_{female}$

        For social class: H_0: $\mu_{upper} = \mu_{middle} - \mu_{lower}$

        For the interaction effect:

        H_0: $\mu_{upper \bullet male} = \mu_{upper \bullet female} = \mu_{middle \bullet male} = \mu_{middle \bullet female} = \mu_{lower \bullet male} = \mu_{lower \bullet female}$

The research hypotheses:

For gender: $H_1: \bar{X}_{male} \neq \bar{X}_{female}$

For social class: $H_1: \bar{X}_{upper} \neq \bar{X}_{middle} \neq \bar{X}_{lower}$

For the interaction effect:

$$\bar{X}_{upper \bullet male} \neq \bar{X}_{upper \bullet female} \neq \bar{X}_{middle \bullet male} \neq \bar{X}_{middle \bullet female} \neq \bar{X}_{lower \bullet male} \neq \bar{X}_{lower \bullet female}$$

2. Set the level of significance: 0.05.

3. Select the appropriate test statistic: a factorial ANOVA.

4. Compute the test statistic value. The SPSS output relating to this test is presented in the following tables:

Between-Subjects Factors

		N
Gender	Female	6
	Male	9
Social_Class	Lower	3
	Middle	5
	Upper	7

Tests of Between-Subjects Effects Dependent Variable: Health_Score

Source	Type III Sum of Squares	df	Mean Square	F	Sig.
Corrected Model	7,623.650[a]	5	1,524.730	3.016	.072
Intercept	35,936.250	1	35,936.250	71.081	.000
Gender	842.917	1	842.917	1.667	.229
Social_Class	5,517.382	2	2,758.691	5.457	.028
Gender * Social_Class	373.808	2	186.904	.370	.701
Error	4,550.083	9	505.565		
Total	64,625.000	15			
Corrected Total	12,173.733	14			

[a] R squared = .626 (adjusted R squared = .419).

5. Determine the value needed for rejection of the null hypothesis: SPSS automatically performs this step, so it does not need to be done manually.

6. Compare the obtained value and the critical value: This step can be completed by looking at the significance values for the main and interaction effects presented in the previous table. As we can see, the main effect of gender is not significant, while the main effect of social class is. Additionally, the interaction between gender and social class is not significant.

7. and 8. Decision time: The null hypothesis that there is no difference in health scores on the basis of gender cannot be rejected, because the main effect is not significant. Additionally, the null hypothesis that there is no interaction effect between gender and social class cannot be rejected, because this interaction effect is also not significant. However, the null hypothesis suggesting no difference in health scores on the basis of social class can be rejected, because this main effect was found to be significant. This suggests that while there is no difference in health scores on the basis of gender and there is no interaction between gender and social class, there is a significant difference in health scores on the basis of social class.

These three results could be written up in the following way, as might be seen in a journal article:

Gender: $F_{(1,9)} = 1.67$, $p = .229$.

Social class: $F_{(2,9)} = 5.46$, $p < .05$.

Interaction between gender and social class: $F_{(2,9)} = 0.37$, $p = .701$.

Between-Subjects Factors

		Value Label	N
Do you work at another job to compensate for your salary?	1.00	No	22
	2.00	Yes, but only during the summer months	33
	3.00	Yes, all year long	15
Degree	1.00	Bachelor's	27
	2.00	Master's	43

Descriptive Statistics

Dependent Variable: Overall Stress

Do you work at another job to compensate for your salary?	Degree	Mean	Std. Deviation	N
No	Bachelor's	2.9286	.53452	7
	Master's	2.5000	.65465	15
	Total	2.6364	.63960	22
Yes, but only during the summer months	Bachelor's	3.3214	.66815	14
	Master's	2.8684	.54879	19
	Total	3.0606	.63440	33
Yes, all year long	Bachelor's	4.0000	.44721	6
	Master's	3.5556	.72648	9
	Total	3.7333	.65101	15
Total	Bachelor's	3.3704	.68770	27
	Master's	2.8837	.72241	43
	Total	3.0714	.74350	70

TESTS OF BETWEEN-SUBJECTS EFFECTS

Dependent Variable: Overall Stress

Source	Type III Sum of Squares	df	Mean Square	F	Sig.
Corrected Model	13.982[a]	5	2.796	7.407	.000
Intercept	601.348	1	601.348	1592.901	.000
WorkAnotherJob	9.380	2	4.690	12.424	.000
Degree	2.876	1	2.876	7.618	.008
WorkAnotherJob * Degree	.002	2	.001	.002	.998
Error	24.161	64	.378		
Total	698.500	70			
Corrected Total	38.143	69			

[a] R squared = .367 (adjusted R squared = .317)

MULTIPLE COMPARISONS

Overall Stress

Bonferroni

(I) Do You Work at Another Job to Compensate for Your Salary?	(J) Do You Work at Another Job to Compensate for Your Salary?	Mean Difference (I-J)	Std. Error	Sig.	95% Confidence Interval	
					Lower Bound	Upper Bound
No	Yes, but only during the summer months	−.4242*	.16911	.044	−.8400	−.0085
	Yes, all year long	−1.0970*	.20574	.000	−1.6028	−.5912
Yes, but only during the summer months	No	.4242*	.16911	.044	.0085	.8400
	Yes, all year long	−.6727*	.19133	.002	−1.1431	−.2023
Yes, all year long	No	1.0970*	.20574	.000	.5912	1.6028
	Yes, but only during the summer months	.6727*	.19133	.002	.2023	1.1431

Based on observed means.
The error term is Mean Square(Error) = .378.

* The mean difference is significant at the 0.05 level.

The test of between-subjects effects shows that there is a main effect of working another job and a main effect of degree on the variable overall stress. There is no interaction between working another job and degree on overall stress levels. The Bonferonni posthoc comparisons show that there is a significant difference in stress between all three levels of working another job. The means in the descriptive statistics show that teachers who don't work any other job are the least likely to agree with the statement "I am very stressed at work," with a mean of 2.64. Teachers who work another job but only during the summer are more likely to agree with the statement than those who don't work any other job (M = 3.06) but they are less likely than those who work another job all year long. Teachers who work another job all year long are the most likely to agree with the statement (M = 3.73).

JUST FOR FUN/CHALLENGE YOURSELF

1. The results are presented in the following tables:

Between-Subjects Factors

		N
Gender	Female	6
	Male	9
Social_Class	Lower	3
	Middle	5
	Upper	7

Multivariate Tests

Effect		Value	F	Hypothesis df	Error df	Sig.
Intercept	Pillai's trace	.892	33.155[a]	2.000	8.000	.000
	Wilks's lambda	.108	33.155[a]	2.000	8.000	.000
	Hotelling's trace	8.289	33.155[a]	2.000	8.000	.000
	Roy's largest root	8.289	33.155[a]	2.000	8.000	.000
Gender	Pillai's trace	.238	1.251[a]	2.000	8.000	.337
	Wilks's lambda	.762	1.251[a]	2.000	8.000	.337
	Hotelling's trace	.313	1.251[a]	2.000	8.000	.337
	Roy's largest root	.313	1.251[a]	2.000	8.000	.337
Social_Class	Pillai's trace	.552	1.717	4.000	18.000	.190
	Wilks's lambda	.449	1.969[a]	4.000	16.000	.148
	Hotelling's trace	1.223	2.140	4.000	14.000	.130
	Roy's largest root	1.220	5.490[b]	2.000	9.000	.028
Gender * Social_Class	Pillai's trace	.286	.751	4.000	18.000	.570
	Wilks's lambda	.727	.690[a]	4.000	16.000	.510
	Hotelling's trace	.356	.623	4.000	14.000	.554
	Roy's largest root	.293	1.319[b]	2.000	9.000	.314

[a] *Exact statistic*

[b] *The statistic is an upper bound on F that yields a lower bound on the significance level.*

[c] *Design:* Intercept + Gender + Social_Class + Gender * Social_Class

Tests of Between-Subjects Effects

Source	Dependent Variable	Type III Sum of Squares	df	Mean Square	F	Sig.
Corrected Model	Lifestyle_Attitudes	7428.567[a]	5	1485.713	3.110	.066
	Health_Score	7623.650[b]	5	1524.730	3.016	.072
Intercept	Lifestyle_Attitudes	34056.010	1	34056.010	71.283	.000
	Health_Score	35936.250	1	35936.250	71.081	.000
Gender	Lifestyle_Attitudes	1254.943	1	1254.943	2.627	.140
	Health_Score	842.917	1	842.917	1.667	.229
Social_Class	Lifestyle_Attitudes	4620.377	2	2310.188	4.835	.037
	Health_Score	5517.382	2	2758.691	5.457	.028
Gender* Social_Class	Lifestyle_Attitudes	289.313	2	144.656	.303	.746
	Health_Score	373.808	2	186.904	.370	.701
Error	Lifestyle_Attitudes	4299.833	9	477.759		
	Health_Score	4550.083	9	505.565		
Total	Lifestyle_Attitudes	60122.000	15			
	Health_Score	64625.000	15			
Corrected Total	Lifestyle_Attitudes	11728.400	14			
	Health_Score	12173.733	14			

[a] R Squared = .633 (Adjusted R Squared = .430)
[b] R Squared = .626 (Adjusted R Squared = .419)

Based on the final table, we see that these results suggest that social class is a significant predictor of both health scores and lifestyle attitudes, while neither gender nor the interaction between gender and social class significantly predicts either of these two dependent variables.

15 Testing Relationships Using the Correlation Coefficient

Cousins or Just Good Friends?

LEARNING OBJECTIVES

- Learn how to test for the significance of a correlation coefficient and how to interpret the results.

- Review the difference between significance and causality in relation to correlation coefficients.

- Review the difference between significance and meaningfulness.

- Learn how to use SPSS to calculate the significance of a correlation coefficient.

SUMMARY/KEY POINTS

- This chapter covers correlation coefficients, which were discussed earlier in the text, but this chapter also covers the use of statistical significance in relation to correlation coefficients.
  - Correlation coefficients examine the relationship between variables, not the difference between groups.
  - A correlation coefficient can only test two variables at a time.
  - The appropriate test statistic to use is the *t*-test for the correlation coefficient.
  - Tests can be either directional or nondirectional. Use a directional (or one-tailed) test when you hypothesize that the relationship will be either positive or negative.

- A significant correlation does not indicate causality.

- A significant correlation does not necessarily indicate a meaningful relationship. A more important determinant of meaningfulness is how large the coefficient of determination is and how small the coefficient of alienation is. The coefficient of determination indicates how much of the variance in one variable is accounted for by the variance in the other variable.

TRUE/FALSE QUESTIONS

1. The correlation coefficient can only be used for two-tailed tests.

2. A significant correlation between two variables does not imply that one variable causes the other.

3. A significant correlation indicates a meaningful relationship exists between the two variables in the analysis.

MULTIPLE-CHOICE QUESTIONS

1. The correlation coefficient examines _____.

   a. differences between two groups

   b. differences between two or more groups

   c. the relationship between two variables

   d. the relationship between two or more variables

2. In the case of the correlation coefficient, the appropriate test statistic to use is the _____.

   a. F-test for the correlation coefficient

   b. t-test for the correlation coefficient

   c. p-test for the correlation coefficient

   d. r-test for the correlation coefficient

3. Which of the following results is significant at the .05 level (two-tailed test)?

   a. $r_{(30)} = .33$

   b. $r_{(60)} = .24$

   c. $r_{(4)} = .79$

   d. $r_{(10)} = .59$

4. Which of the following results is significant at the .01 level (one-tailed test)?

   a. $r_{(30)} = .42$

   b. $r_{(10)} = .62$

   c. $r_{(4)} = .87$

   d. $r_{(5)} = .81$

SHORT-ANSWER/ESSAY QUESTIONS

1. What is the critical value of the correlation coefficient needed for rejection of the null hypothesis if your degrees of freedom is 15 and you are conducting a one-tailed test using the .01 level of significance?

2. What is the critical value of the correlation coefficient needed for rejection of the null hypothesis if your degrees of freedom is 30 and you are conducting a two-tailed test using the .05 level of significance?

3. If your total sample size is 75 paraprofessionals, what is your degrees of freedom for the correlation coefficient?

4. Your degrees of freedom is 55. Looking at the table of critical values for the correlation coefficient, you can see that there are only entries for 50 and 60 degrees of freedom, not 55. If you want to be more conservative, which entry should you choose?

5. How would you interpret $r_{(37)} = .89, p < .05$?

6. In your research you found that the relationship between student motivation and student achievement in math was $r_{(189)} = .43, p < .05$. How would you explain this when writing up your results?

SPSS QUESTION

1. Using SPSS, perform the eight steps of hypothesis testing to test the null hypothesis that there is no relationship between IQ and salary, using the following set of data. Also, report your result in the following form: $r_{(17)} = .55, p > .05$.

Salary ($1,000)	IQ
245	127
120	133
90	98
88	105
75	115
74	102
66	115
58	98
45	80
23	85
21	78
15	70

2. Open the supplemental data set "Teacher Survey Data" in SPSS. Use SPSS to calculate the correlation coefficient for the relationship between "School Administration" and "Overall Satisfaction" levels in teachers. What does this correlation tell you about the relationship between school administration and the overall satisfaction of teachers?

JUST FOR FUN/CHALLENGE YOURSELF

1. Why do the critical values for the correlation coefficient decrease as the sample size increases? Why do they increase when you have a lower value for the possibility of Type I error?

ANSWER KEY

TRUE/FALSE QUESTIONS

1. False. The correlation coefficient can be used for both one-tailed and two-tailed tests.

2. True.

3. False. You can have a significant correlation that is very weak, meaning that there is a very weak (hence, not meaningful) relationship between the two variables.

MULTIPLE-CHOICE QUESTIONS

1. (c) the relationship between two variables

2. (b) t-test for the correlation coefficient

3. (d) $r_{(10)} = .59$

4. (a) $r_{(30)} = .42$

SHORT-ANSWER/ESSAY QUESTIONS

1. The critical value is .5577.

2. The critical value is .3494.

3. The degrees of freedom $= n - 2 = 75 - 2 = 73$.

4. To be more conservative, you choose the entry for 50 degrees of freedom, which results in a higher critical value for the correlation coefficient. This in turn makes the test more conservative, with a finding of significance less likely.

5. First, r represents the test statistic that was used, 37 is the number of degrees of freedom, and .89 represents the obtained value that was calculated for the correlation coefficient. Finally, $p < .05$ indicates the probability is less than 5% that the relationship between the two variables is due to chance alone. We can conclude that there is a significant relationship between the two variables.

6. A statistically significant correlation of .43 between student motivation and student math achievement is a moderate, positive relationship. Thus, the more motivated a student is the more likely he or she is to achieve at higher levels in math.

SPSS QUESTION

1. The eight steps of hypothesis testing:

   1. State the null and research hypotheses:

      $H_0: \rho_{xy} = 0$

      $H_1: r_{xy} \neq 0$

   2. Set the level of significance: .05.

   3. Select the appropriate statistic: the correlation coefficient.

   4. Compute the test statistic value: This is done for us automatically in SPSS. The following table illustrates the SPSS output.

Correlations

		Salary_1k	IQ
Salary_1k	Pearson Correlation	1	0.772**
	Sig. (2-tailed)		0.003
	N	12	12
IQ	Pearson Correlation	0.772**	1
	Sig. (2-tailed)	0.003	
	N	12	12

** Correlation is significant at the 0.01 level (2-tailed).

5. Determine the value needed for rejection of the null hypothesis: Because our sample size is 12, we know that our degrees of freedom is 10. As we are conducting a two-tailed test using the .05 level of significance, our critical value is .5760. Note: This step does not necessarily need to be done, because SPSS automatically calculates the significance level for the correlation coefficient.

6. Compare the obtained value with the critical value: Our obtained value is higher than our critical value.

7. and 8. Decision time: Because our obtained value is higher than our critical value, you can reject the null hypothesis that there is no relationship between IQ and salary. We obtained a positive correlation coefficient, indicating that there is a positive association between these two variables; in other words, higher IQ is associated with higher salary.

Additionally, this result can be reported as: $r_{(10)} = .772$, $p < .05$.

SPSS QUESTION 2

The correlation is below:

Correlations

		School Administration	Overall Satisfaction
School Administration	Pearson Correlation	1	−.056
	Sig. (2-tailed)		.644
	N	70	70
Overall Satisfaction	Pearson Correlation	−.056	1
	Sig. (2-tailed)	.644	
	N	70	70

This correlation is not significant, meaning that there is no relationship between school administration (how supportive and encouraging administration is and how often the principal communicates with teachers) and teachers' overall satisfaction.

JUST FOR FUN/CHALLENGE YOURSELF

1. Critical values for the correlation coefficient (more generally) decrease as the sample size increases because, as we have a greater number of individuals, it is easier for a relationship of a certain strength to be seen; hence, when you have a lot of data, you don't need a very strong relationship to achieve statistical significance.

   On the other hand, when your sample size is very small, the correlation coefficient must be very high for you to be able to say that the relationship is valid and not just due to chance. Additionally, a lower value for the probability of Type I error means that you have a higher degree of certainty that the relationship is valid and not due to chance. Therefore, your correlation coefficient must be that much higher to support a greater degree of certainty.

16 Using Linear Regression

Predicting the Future

LEARNING OBJECTIVES

- Learn about linear regression and how it can be used for prediction.

- Understand when it is appropriate to use linear regression.

- Learn how to determine the accuracy of your predictions.

- Understand when it is and when it is not appropriate to include multiple independent variables in what is called multiple regression.

SUMMARY/KEY POINTS

- Linear regression, in essence, uses correlations between variables as the basis for predicting the value of one variable based on the value of another.
  - The higher the absolute value of the correlation coefficient, the more accurate the prediction. In the case of linear regression, a perfect correlation translates into a perfect prediction.
  - In linear regression, a regression equation is determined, and this equation can be used to plot a regression line. The regression line reflects the best estimate of predicted scores for the dependent variable based on levels of the independent variable.
    - In the regression equation, the predicted score of the dependent variable is equal to the slope multiplied by the value of the independent variable, plus a constant that is equal to the point at which the regression line crosses the y-axis.
  - Error in prediction (error of estimate) is calculated as the distance between each individual data point and the regression line—the error value indicates by how much your prediction was "off."
    - Standard error of estimate is the average of all values for error in prediction. This value tells you how imprecise the predictive power of the linear regression analysis is overall.
    - Error decreases as the correlation between the two variables increases.

- In linear regression, the outcome variable is called the criterion or dependent variable and written as Y, while the predictor or independent variable is written as X.

- In multiple regression, more than one independent variable is included in the analysis.
  - An additional independent variable should only be included if it makes a unique contribution to the understanding, or prediction, of the dependent variable.
  - Additionally, multiple independent variables are best included in a regression analysis if they are uncorrelated with one another but are all correlated with the dependent variable.

KEY TERMS

- **Regression equation**: In regression, an equation that defines the line that has the best fit with your data

- **Regression line**: The line of best fit that is drawn (or calculated) based on the values in the regression equation

- **Line of best fit**: The regression line that best fits the data and minimizes the error in prediction

- **Error in prediction**: (aka error of estimate): The difference between the actual score and the predicted score in a regression

- **Criterion or dependent variable**: The outcome variable, or the variable that is predicted, in a regression analysis

- **Predictor or independent variable**: The variable that is used to predict the dependent variable in a regression analysis

- **Y prime**: The predicted value of Y, the dependent variable, written as Y'

- **Standard error of estimate**: The average amount that each data point differs from the predicted data point

- **Multiple regression**: A type of regression in which more than one independent variable is included in the analysis

TRUE/FALSE QUESTIONS

1. Linear regression uses correlations as its basis.

2. Linear regression can be used to predict values of the dependent variable for individuals outside of your data set.

3. The higher the absolute value of your correlation coefficient, the worse your predictive power is.

4. When using multiple regression, it is always best to include as many predictor variables as possible.

5. When using multiple regression, it is best to select independent variables that are uncorrelated with one another but are all related to the predicted variable.

MULTIPLE-CHOICE QUESTIONS

1. Your prediction in linear regression will be perfect if your correlation is _____.

   a. −1

   b. +1

   c. 0

   d. either a or b

2. If the prediction in a linear regression analysis were perfect, all predicted points would fall _____.

   a. on the regression line

   b. above the regression line

   c. below the regression line

   d. both above and below the regression line

3. If your standard error of estimate is high, the plot of the regression line with data points will show _____.

   a. data points very close to the regression line

   b. data points very far from the regression line

   c. data points exactly on the regression line

   d. any of a, b, or c—it cannot be determined

4. You measure the quality of students' relationships with their teacher and use that to predict how engaged students will be in class. The variable of how engaged students are in class is _____.

   a. the criterion variable

   b. the dependent variable

   c. the predictor variable

   d. the independent variable

   e. both a and b

5. In the study presented in question 4, the variable of quality of students' relationships with their teacher is the _____.

   a. criterion variable

   b. dependent variable

   c. predicted variable

   d. independent variable

6. In linear regression, the dependent variable is indicated by which of the following?

   a. Y

   b. b

   c. X

   d. a

7. In linear regression, the independent variable is indicated by which of the following?

   a. Y

   b. b

   c. X

   d. a

8. In linear regression, the slope is indicated by which of the following?

   a. Y

   b. b

   c. X

   d. a

9. In linear regression, the point at which the line crosses the y-axis is indicated by which of the following?

   a. Y

   b. b

   c. X

   d. a

10. The predicted value of *Y* is called _____.

    a. *Y* prime

    b. *Y* sigma

    c. *Y* delta

    d. *Y* alpha

11. In linear regression, the regression line can be _____.

    a. a straight line only

    b. a straight or curved line

    c. a curved line only

    d. none of the above

12. If you have a correlation coefficient of –1, the standard error of estimate will be equal to which of the following?

    a. 1

    b. –1

    c. 2

    d. 0

EXERCISES

1. Using the following data, conduct a linear regression analysis (by hand). In this analysis, you are testing whether SAT scores are predictive of overall college GPA. Also, write out the regression equation. How would you interpret your results?

SAT Score	GPA
670	1.2
720	1.8
750	2.3
845	1.9
960	3.0
1,000	3.3
1,180	3.2
1,200	3.4
1,370	2.9
1,450	3.8
1,580	4.0
1,600	3.9

2. You conduct a linear regression in which the number of hours per week spent playing outside is used to predict students' overall general health, measured on a scale from 0 to 100

with 100 being the best possible health. You obtain the following results: $b = 4.5$; $a = 37$. Using the general formula for a regression line, calculate the predicted values for overall general health for students who spend 0 hours per week playing outside, 3 hours per week playing outside, and 12 hours per week playing outside.

SHORT-ANSWER/ESSAY QUESTIONS

1. When you are deciding which variables to include as predictors in a multiple regression equation, what are some conditions that you must consider first?

2. Describe how correlation and regression are linked yet distinct.

SPSS QUESTION

1. Using the data presented under the "Exercises" section, question 1, conduct the same linear regression using SPSS and interpret the output. Additionally, create a scatterplot of the data with a superimposed regression line.

2. Open the supplemental data set "Teacher Survey Data" in SPSS. Use SPSS to conduct a linear regression to test whether teacher type predicts overall stress. Create a scatterplot for this data and then interpret the results.

JUST FOR FUN/CHALLENGE YOURSELF

1. Starting with the data from the "Exercises" section, question 2, you find two new individuals who have health scores of 100 and 75, respectively. Calculate the predicted values for the number of hours they exercise per week.

2. Using the following data, conduct a multiple linear regression in SPSS and interpret the results. SAT score and IQ are both independent variables, and GRE score is the dependent variable. Interpret the results and write out the regression equation.

SAT Score	IQ	GPA
670	80	1.2
720	87	1.8
750	105	2.3
845	95	1.9
960	110	3.0
1,000	98	3.3
1,180	110	3.2
1,200	125	3.4
1,370	115	2.9
1,450	120	3.8
1,580	140	4.0
1,600	135	3.9

TRUE/FALSE QUESTIONS

1. True.

2. True.

3. False. The higher the absolute value of your correlation coefficient, the better your predictive power is.

4. False. Careful judgment should be used when deciding which predictor variables to include.

5. True.

MULTIPLE-CHOICE QUESTIONS

1. (d) either a or b

2. (a) on the regression line

3. (b) data points very far from the regression line

4. (e) both a and b

5. (d) independent variable

6. (a) Y

7. (c) X

8. (b) b

9. (d) a

10. (a) Y prime

11. (a) a straight line only

12. (d) 0

EXERCISES

1. The regression coefficients are calculated using the following equations:

$$b = \frac{\sum XY - \left(\sum X \sum y / n\right)}{\sum X^2 - \left[\left(\sum X\right)^2 / n\right]} \Rightarrow$$

$$\frac{41,509.5 - (13,325 \times 34.7 / 12)}{16,033,625 - 13,325^2 / 12}$$

$$a = \frac{\sum Y - b \sum X}{n} \Rightarrow$$

$$\frac{34.7 - 0.002 \times 13,325}{12} = 0.219$$

The regression equation would be represented as follows:

$$Y' = 0.002X + 0.219$$

The value for a, 0.219, indicates the predicted GPA if the SAT score were equal to zero. The value for b, 0.002, indicates that a 1-unit increase in SAT score is associated with a 0.002 predicted increase in GPA. Multiplying this figure by 100, we can say that a 100-unit increase in SAT score is associated with a 0.2 predicted increase in GPA. This means, more generally, that higher SAT scores are associated with higher GPAs.

2. Here is the regression equation:

$$Y' = 4.5X + 37$$

For students who spend 0 hours per week playing outside:

$$Y' = 4.5(0) + 37 = 37$$

For students who spend 3 hours per week playing outside:

$$Y' = 4.5(3) + 37 = 40.5$$

For students who spend 12 hours per week playing outside:

$$Y' = 4.5(12) + 37 = 81$$

SHORT-ANSWER/ESSAY QUESTIONS

1. In deciding which predictor variables to add into a regression equation, keep in mind a number of considerations. You must think about whether a new variable will make a unique contribution to understanding the dependent variable. The two (or more) variables in combination should predict Y better than any of the variables do alone. You must also balance the cost of resources needed for another predictor with the possible benefit of adding another predictive value. Additionally, there must be a theoretical (or literature-based) rationale for adding another predictor; there is a limit to how many variables will theoretically contribute to a prediction. Finally, when including multiple independent/predictor variables in a regression analysis, it is best if they are uncorrelated with one another but are both correlated to the dependent variable.

2. Correlation is used to report only the relationship between two variables, whereas in regression, one variable is designated as the independent/predictor variable and the other variable as the dependent/criterion variable. This designation allows regression to be used to predict the value of one variable from the value of another. Correlation is used in the steps of conducting a regression. In fact, linear regression uses correlations between variables as the basis for predicting the value of one variable based on the value of another. In linear regression, a perfect correlation would translate into perfect prediction. In multiple regression, correlations are run as a preliminary step to assure the investigator that the independent/predictor variables are not strongly correlated with one another and to determine that each independent/predictor variable is correlated with the dependent/criterion variable. Finally, neither correlation nor regression results can be used to claim causality between variables.

1. The SPSS output is shown here:

Model Summary

Model	R	R Square	Adjusted R Square	Std. Error of the Estimate
dimension0 1	.893[a]	.797	.777	.42679

[a] Predictors: (Constant), SAT

ANOVA[b]

Model	Sum of Squares	df	Mean Square	F	Sig.
Regression	7.168	1	7.168	39.351	.000[a]
	1.821	10	.182		
Residual Total	8.989	11			

[a] Predictors: (Constant), SAT
[b] Dependent Variable: GRE

Coefficients[a]

Model		Unstandardized Coefficients		Standardized Coefficients		
		B	Std. Error	Beta	t	Sig.
1	(Constant)	.219	.444		.494	.632
	SAT	.002	.000	.893	6.273	.000

[a] Dependent Variable: GPA

The final table, "Coefficients," gives us our a and b values.

The value for a, 0.219, indicates the predicted GPA if the SAT score were equal to zero. The value for b, 0.002, indicates that a 1-unit increase in SAT score is associated with a 0.002 predicted increase in GPA. Multiplying this figure by 100, we can say that a 100-unit increase in SAT score is associated with a 0.2 predicted increase in GPA. This means, more generally, that higher SAT scores are associated with higher GPAs. Additionally, the effect of SAT was found to be significant at the .05 level of significance.

The scatterplot with superimposed regression line is shown here:

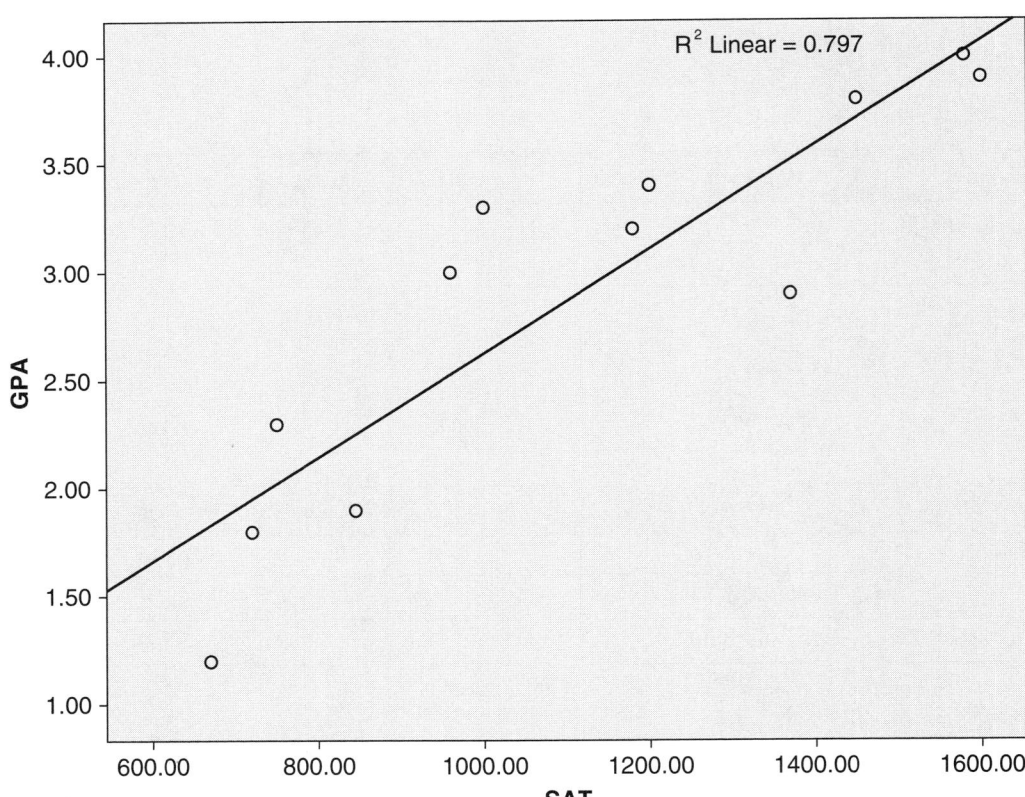

2. The SPSS output is shown here:

Model	R	R Square	Adjusted R Square	Std. Error of the Estimate
1	.014[a]	.000	−.015	.74888

[a] *Predictors:* (Constant), Teacher Type

ANOVA[a]

	Model	Sum of Squares	df	Mean Square	F	Sig.
1	Regression	.007	1	.007	.013	.910[b]
	Residual	38.136	68	.561		
	Total	38.143	69			

[a] *Dependent Variable:* Overall Stress
[b] *Predictors:* (Constant), Teacher Type

Coefficients[a]

| | Model | Unstandardized Coefficients | | Standardized Coefficients | | |
		B	Std. Error	Beta	t	Sig.
1	(Constant)	3.057	.153		19.958	.000
	TeacherType	.005	.040	.014	.113	.910

[a] *Dependent Variable:* Overall Stress

The final table, "Coefficients," gives us our *a* and *b* values.

The value for *a*, 3.057, indicates teachers' reported stress if years of experience were equal to zero. The value for *b*, 0.005, indicates that a 1-unit increase in years of experience is associated with a 0.005 predicted increase in stress. Overall, years of experience was not found to be a significant predictor of reported teacher stress with p >.05.

The scatterplot is shown here:

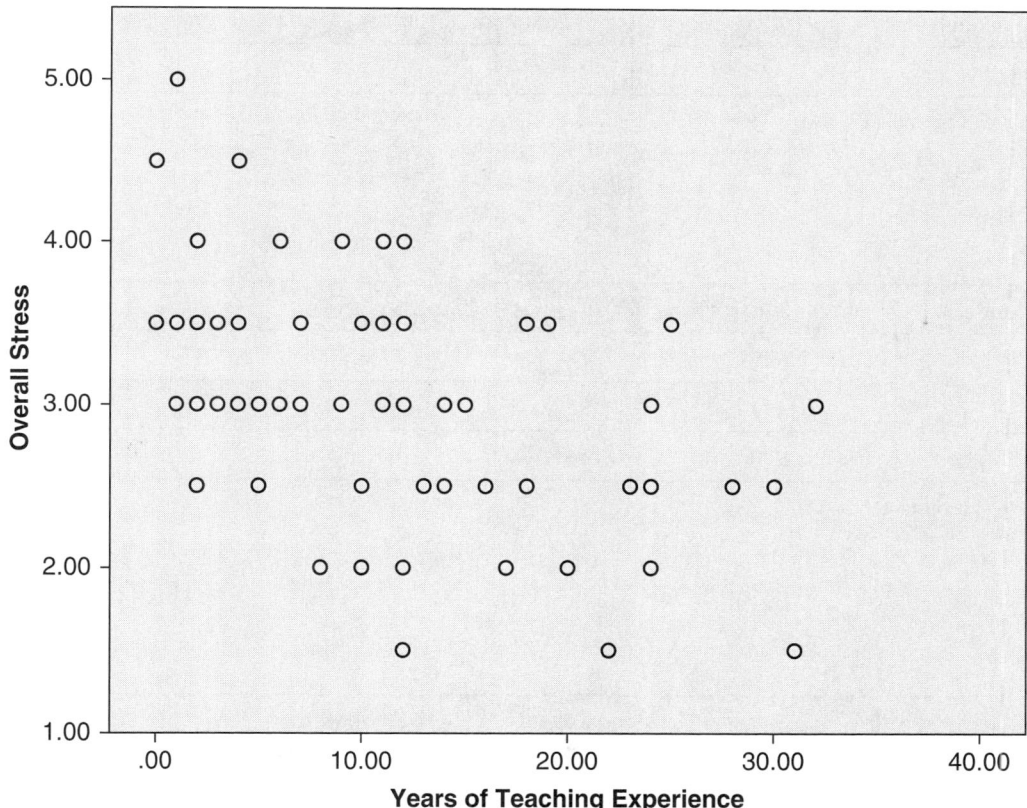

JUST FOR FUN/CHALLENGE YOURSELF

1. The results were *b* = 4.5 and *a* = 37. These values give us this equation:

$$Y' = 4.5X + 37$$

For the respondent with a health score of 100,

$$Y' = 4.5X + 37 \Rightarrow$$
$$100 = 4.5X + 37$$
$$63 = 4.5X$$
$$X = 14$$

For the respondent with a health score of 75,

$$Y' = 4.5X + 37 \Rightarrow$$
$$75 = 4.5X + 37$$
$$38 = 4.5X$$
$$X = 8.44$$

2. The SPSS output is shown here:

Model Summary

Model	R	R Square	Adjusted R Square	Std. Error of the Estimate
1	.917[a]	.841	.805	.39901

[a] Predictors: (Constant), IQ, SAT

ANOVA[b]

Model	Sum of Squares	df	Mean Square	F	Sig.
Regression	7.556	2	3.778	23.731	.000[a]
	1.433	9	.159		
Residual Total	8.989	11			

[a] Predictors: (Constant), IQ, SAT
[b] Dependent Variable: GPA

Coefficients[a]

Model		Unstandardized Coefficients		Standardized Coefficients		
		B	Std. Error	Beta	t	Sig.
11	(Constant)	21.144	.966		21.184	.267
	SAT	.001	.001	.432	1.334	.215
	IQ	.025	.016	.506	1.562	.153

[a] Dependent Variable: GPA

Neither SAT nor IQ were found to be significant. However, based on the coefficients, a 1-unit increase in SAT is associated with a 0.001-unit increase in GPA, while a 1-unit increase in IQ is associated with a 0.025-unit increase in GPA. The value for a, -1.144, indicates the predicted GPA if the SAT score and IQ were equal to zero.

Here is the regression equation: $Y' = 0.001(SAT) + 0.025(IQ) - 1.144$.

17 Chi-Square and Some Other Nonparametric Tests

What to Do When You're Not Normal

LEARNING OBJECTIVES

- Understand the reason behind the use of nonparametric statistics and when they should be preferred.

- Learn about chi-square and how it is calculated, both by hand and by using SPSS.

- Briefly review some other nonparametric statistics.

SUMMARY/KEY POINTS

- The statistical tests covered previously in this book consisted of parametric statistical tests, which make certain assumptions about the data, sample size, etc.

- Nonparametric statistics do not incorporate these same assumptions. These types of statistics can be used when the assumptions of parametric statistical tests have been violated.
  - Nonparametric statistics can be preferred when you have sample sizes that are very small (less than 30) or are analyzing categorical variables.

- The nonparametric test highlighted in this chapter is chi-square.
  - A one-sample chi-square focuses on the distribution of a single variable. It is used to determine whether the distribution of a single categorical variable is significantly different from what would be expected by chance.
    - The one-sample chi-square test is also called goodness of fit.
  - A two-sample chi-square focuses on the relationship between two categorical variables. It determines whether the variables are significantly related, or dependent.
    - The two-sample chi-square test is also called the test of independence.
  - The chi-square statistic measures the difference between the observed data and what would be expected by chance alone.

- The chapter also lists nine other nonparametric tests, when the tests should be used, and sample research questions for each.

KEY TERMS

- **Parametric statistics**: A set of statistical tests that incorporate certain assumptions, often including, among others, a sufficient sample size and a normal or near-normal distribution of continuous variables

- **Nonparametric statistics** (aka distribution-free statistics): A set of statistical tests that do not incorporate the assumptions held by parametric statistical tests

- **Chi-square**: A nonparametric statistical test used to determine whether a single variable has a distribution that would be expected through chance or whether a general relationship exists between two categorical variables

TRUE/FALSE QUESTIONS

1. Nonparametric statistics include more assumptions than do parametric statistics.

2. Chi-square is a nonparametric statistical test.

MULTIPLE-CHOICE QUESTIONS

1. Which of the following tests is also known as goodness of fit?

   a. One-sample chi-square

   b. Two-sample chi-square

   c. Any nonparametric statistical test

   d. Fisher's exact test

2. Which of the following tests is used to determine whether the distribution of a single categorical variable significantly differs from what would be expected by chance?

   a. Two-sample chi-square

   b. Fisher's exact test

   c. Kolmogorov-Smirnov test

   d. One-sample chi-square

3. Which of the following tests is used to determine whether two categorical variables are related, or dependent?

   a. Two-sample chi-square

   b. The sign test

   c. Kolmogorov-Smirnov test

   d. One-sample chi-square

4. Nonparametric statistics might be preferred under which of the following conditions?

   a. Your sample size is very small.

   b. Your sample size is very large.

   c. The variables you are analyzing are continuous.

   d. All the assumptions of parametric statistics have been met.

5. If there is no difference between what is observed and what is expected, your chi-square value will be _____.

   a. 1

   b. 0

   c. −1

   d. a, b, or c—it is impossible to determine

6. Which of the following values is significant at the .05 level?

   a. $\chi^2_{(10)} = 15.43$.

   b. $\chi^2_{(8)} = 12.20$.

   c. $\chi^2_{(5)} = 11.12$.

   d. $\chi^2_{(2)} = 4.23$.

7. Which of the following values is significant at the .01 level?

   a. $\chi^2_{(12)} = 22.14$.

   b. $\chi^2_{(6)} = 14.20$.

   c. $\chi^2_{(4)} = 8.42$.

   d. $\chi^2_{(2)} = 9.45$.

EXERCISE

1. Using the following data, perform the eight steps of hypothesis testing. In this question, calculate the chi-square statistic by hand.

Expected Number	Observed Number
10	15
20	15
20	37
50	33

SHORT-ANSWER/ESSAY QUESTION

1. Interpret the following: $\chi^2_{(3)} = 14.2, p < .05$.

SPSS QUESTION

1. Using the following data, perform a chi-square analysis using SPSS. Interpret your results.

Expected Number	Observed Number
10	8
10	10
10	11
10	9

2. Open the supplemental data set "Teacher Survey Data" in SPSS. Use SPSS to perform a chi-square examining the relationship between grade level taught and degree.

JUST FOR FUN/CHALLENGE YOURSELF

1. What test can be used to see whether scores from a sample come from a specified population?

2. Which test computes the exact probability of outcomes in a 2 × 2 table?

3. Which test computes the correlation between ranks?

ANSWER KEY

TRUE/FALSE QUESTIONS

1. False. Nonparametric statistics include fewer assumptions than do parametric statistics.

2. True.

MULTIPLE-CHOICE QUESTIONS

1. (a) One-sample chi-square

2. (d) One-sample chi-square

3. (a) Two-sample chi-square

4. (a) Your sample size is very small.

5. (b) 0

6. (c) $\chi^2_{(5)} = 11.12$.

7. (d) $\chi^2_{(2)} = 9.45$.

EXERCISE

1. The eight steps of hypothesis testing:

   1. State the null and research hypotheses:

$$H_0: P_1 = P_2 = P_3 = P_4$$
$$H_1: P_1 \neq P_2 \neq P_3 \neq P_4$$

   2. Set the level of significance: 0.05.

   3. Select the appropriate test statistic: chi-square.

   4. Compute the test statistic value:

Expected Number	Observed Number
10	15
20	15
20	37
50	33

$$\chi^2 = \sum \frac{(O-E)^2}{E} \Rightarrow$$

$$\frac{(15-10)^2}{10} + \frac{(15-20)^2}{20} + \frac{(37-20)^2}{20} + \frac{(33-50)^2}{50} = 23.98$$

5. Determine the value needed to reject the null hypothesis: Critical $\chi^2_{(3)} = 7.82$.

6. Compare the obtained value with the critical value: The obtained value, 23.98, is larger than the critical value of 7.82.

7. and 8. Decision time: Because the obtained value is greater than the critical value, we reject the null hypothesis that there is no difference between these groups.

SHORT-ANSWER/ESSAY QUESTION

1. First, χ^2 (chi-square) represents the test statistic. The value of 3 represents the degrees of freedom, while 14.2 is the obtained value arrived at by using the formula for chi-square. Finally, $p < .05$ indicates that the probability is less than 5% that the variable is equally distributed across all categories.

SPSS QUESTION

1. The SPSS output is shown in the following table. The results indicate that this test did not find statistical significance, meaning that the distribution of the variable cannot be said to differ significantly from that expected through chance alone.

Test Statistics

	Var1
Chi-square	.600[a]
Df	3
Asymp. Sig.	.896

[a] *Zero cells (0.0%) have expected frequencies less than 5. The minimum expected cell frequency is 10.0.*

2. The SPSS output is shown in the following table. The results do not show statistical significance, thus, the number of teachers with master's degrees at each grade level do not differ significantly from what is expected based on chance alone.

Grade Level Taught * Degree Crosstabulation				
		Degree		
Count		**Bachelor's**	**Master's**	**Total**
Grade Level Taught	Kindergarten	1	5	6
	First	3	5	8
	Second	5	3	8
	Third	3	3	6
	Fourth	2	4	6
	Fifth	4	3	7
	Multiple Grade Levels	9	20	29
Total		27	43	70

Chi-Square Tests			
	Value	***df***	**Asymp. Sig. (2-sided)**
Pearson Chi-Square	5.267[a]	6	.510
Likelihood Ratio	5.334	6	.502
Linear-by-Linear Association	.054	1	.816
N of Valid Cases	70		

[a] 12 cells (85.7%) have expected count less than 5. The minimum expected count is 2.31.

JUST FOR FUN/CHALLENGE YOURSELF

1. The Kolmogorov-Smirnov test

2. Fisher's exact test

3. The Spearman rank correlation coefficient

18 Some Other (Important) Statistical Procedures You Should Know About

- Review some more advanced statistical procedures and when and how they are used.

SUMMARY/KEY POINTS

- This chapter presents an overview of some more advanced statistical tests.

- MANOVA (multiple analysis of variance) is used when you want to include more than one dependent variable in an analysis. This can be preferred when the dependent variables are related to one another, because it is difficult to determine the effect of the treatment or independent variable on any one outcome. A MANOVA can be used to examine effects of a summer school on multiple dependent variables such as student achievement, motivation, and school retention.

- Repeated measures analysis of variance is used when participants are tested more than once on one factor. This is often used in education when we measure student growth from the beginning of the school year to the end.

- Analysis of covariance is used to control for the effect of one or more continuous variables. This can be used to control for the effects of socioeconomic status or intelligence levels when examining students' response to an educational intervention.

- Multiple regression is a type of linear regression that includes more than one independent variable. This can be used to examine multiple factors that predict teacher retention rates.

- Meta-analysis involves combining data from several studies so you can examine patterns and trends. This can be used to examine multiple educational interventions to determine which intervention techniques yield the greatest effect sizes.

- Discriminant analysis is used to see what sets of variables discriminate between two sets of individuals.

- Factor analysis is used to determine how closely related a set of items is and how the items might form clusters or factors. Factor analysis can be preferred because factors can be better than individual variables at representing outcomes. Factor analysis is a common technique used when creating a new assessment.

- Path analysis studies causality between variables.

- Structural equation modeling (SEM) is an extension of path analysis and is also considered a generalization of regression and factor analysis. SEM is considered confirmatory as opposed to exploratory.

19 Data Mining

An Introduction to Getting the Most Out of Your BIG Data

CHAPTER OUTLINE

- ✦ Our Sample Data Set—Who Doesn't Love Babies?
- ✦ Counting Outcomes
  - ✧ Counting With Frequencies
- ✦ Pivot Tables and Cross-Tabulation: Finding Hidden Patterns
  - ✧ Creating a Pivot Table
  - ✧ Modifying a Pivot Table
- ✦ Summary
- ✦ Time to Practice

LEARNING OBJECTIVES

- Understand what data mining is.

- Understand how data mining is used to make sense out of very big data sets.

- Understand how to use SPSS to apply basic data mining tools.

- Apply pivot tables to the analysis of large data sets.

SUMMARY/KEY POINTS

- This chapter covers the analysis of big data, such as health care records, social media interactions, customers' purchasing patterns, and daily physical activity.
  - Data mining is frequently used in sales and marketing, and is also called analytics.
  - When you consider using a big data set, keep in mind the quality of the data just as you would with other data sets.
  - Big data is a very large collection of either case or variables or, often, both.
  - Big data sets can be pulled from the Internet. Because of this, new data are added when they become available. If you use it at different times, the answers will be different so expect that.

- You can do basic counts in SPSS of big data with frequencies. The path begins with Analyze—Descriptive Statistics—Frequencies, and can be done with numeric variables or text/string variables.

- You can also do basic counts in SPSS of big data with crosstabs to summarize categorical data. The path begins with Analyze—Descriptive Statistics—Crosstabs.

- You can use pivot tables—with frequencies, crosstabs, or other output—in SPSS to further mine your data.

KEY TERMS

- **Exabytes**: About 1 quintillion bytes; introduced in this chapter to alert readers to how very large the possibilities are for data mining

- **Data mining**: Looking for patterns in large data sets; also called analytics

- **Cross-tabulation tables**: Known as crosstabs, summarize categorical data to create column and row totals and cell counts

- **Pivot table**: Table that allows you to easily visualize and manipulate rows and columns, as well as the cells' contents

TRUE/FALSE QUESTIONS

1. Data mining is used by online stores to recommend future purchases based on your current purchases.

2. When using big data sets from Internet sources, expect the the data within to be stable from month to month.

3. Procedures for mining Big Data can only be used for big data sets.

SPSS QUESTIONS

Use the British Election Study data set entitled Version 2.2 Face-to-face Post-election Survey (with vote validation). It is dated May 2015–September 2015.

www.britishelectionstudy.com/data-objects/cross-sectional-data/

Use this data set to create frequencies and crosstabs on select variables in order to make sense of this data set, which is very large.

1. Choose the following two variables: m02_1 *Politicians don't care what people like me think*; and, b02 *Which party did you vote for in the general election?* First, create a frequency table for each of them.

2. Next, run crosstabs on both variables.

3. Next, create a pivot table from the crosstabs results.

4. and 5. Finally, go to Graphs to create two cluster bars. You may follow the instructions in the textbook, or you may choose to use Graphs—Legacy Dialogs—Bar—Bar Charts—Cluster. Create the cluster bar both ways: with m02_1 on the *x*-axis and b02 on the *y*-axis, and then b02 on the *x*-axis and m02_1 on the *y*-axis.

JUST FOR FUN/CHALLENGE YOURSELF

1. After you create both cluster bars, give reasons why each one would be the best way to present the data.

ANSWER KEY

TRUE/FALSE QUESTIONS

1. True.

2. False. With the acquisition of new data, many data sets on the Internet are updated, so expect different numbers if you utilize the same link to data at a later time.

3. False. All the same procedures can also be used to make sense of small data sets.

SPSS QUESTION

1. Frequency table for m02_1 *Politicians don't care about what people like me think.*

Politicians don't care what people like me think		Frequency	Percent	Valid Percent	Cumulative Percent
Valid	Don't know	30	1.0	1.0	1.0
	Strongly disagree	89	3.0	3.0	4.0
	Disagree	762	25.5	25.5	29.5
	Neither agree nor disagree	578	19.4	19.4	48.8
	Agree	1029	34.4	34.4	83.3
	Strongly agree	499	16.7	16.7	100.0
	Total	2987	100.0	100.0	

Frequency table for b02 *Which party did you vote for in the general election?*

Which party did you vote for in the general election?		Frequency	Percent	Valid Percent	Cumulative Percent
Valid	Refused	89	3.0	4.1	4.1
	Don't know	11	.4	.5	4.6
	Labour	677	22.7	30.8	35.4
	Conservatives	836	28.0	38.1	73.4
	Liberal Democrats	158	5.3	7.2	80.6
	Scottish National Party (SNP)	101	3.4	4.6	85.2
	Plaid Cymru	10	.3	.5	85.7
	Green Party	66	2.2	3.0	88.7
	United Kingdom Independence Party (UKIP)	234	7.8	10.7	99.3
	British National Party (BNP)	3	.1	.1	99.5
	Other	12	.4	.5	100.0
	Total	2197	73.6	100.0	
Missing	System	790	26.4		
Total		2987	100.0		

2. Crosstabs for both variables

Politicians don't care what people like me think * B2 Which party did you vote for in the general election? Crosstabulation

Count

		B2 Which party did you vote for in the general election?											Total
		Refused	Don't know	Labour	Conservatives	Liberal Democrats	Scottish National Party (SNP)	Plaid Cymru	Green Party	United Kingdom Independence Party (UKIP)	British National Party (BNP)	Other	
Politicians don't care what people like me think	Don't know	1	1	6	2	0	1	0	0	2	0	0	13
	Strongly disagree	3	0	16	51	3	1	0	2	2	0	1	79
	Disagree	21	1	144	350	55	24	5	17	29	1	2	649
	Neither agree nor disagree	16	2	131	172	37	18	1	15	41	1	1	435
	Agree	32	6	250	214	50	39	4	24	94	0	5	718
	Strongly agree	16	1	130	47	13	18	0	8	66	1	3	303
Total		89	11	677	836	158	101	10	66	234	3	12	2197

3. Pivot table for both variables

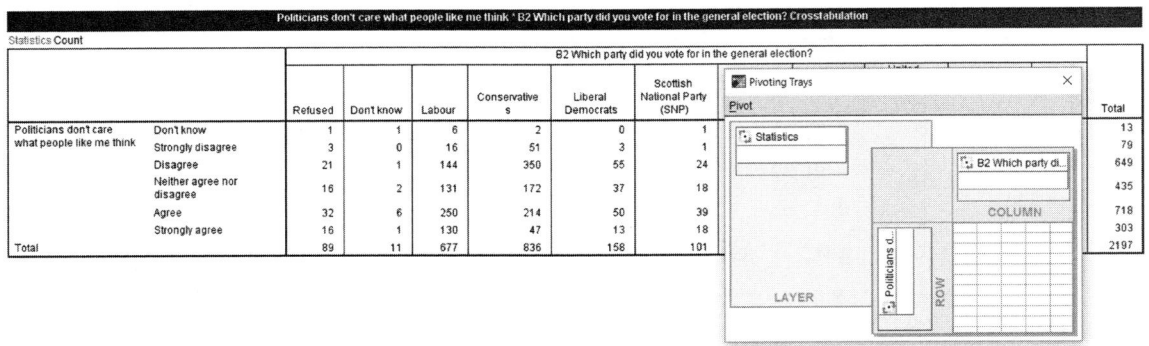

4. m02_1 *Politicians don't care about what people like me think.*

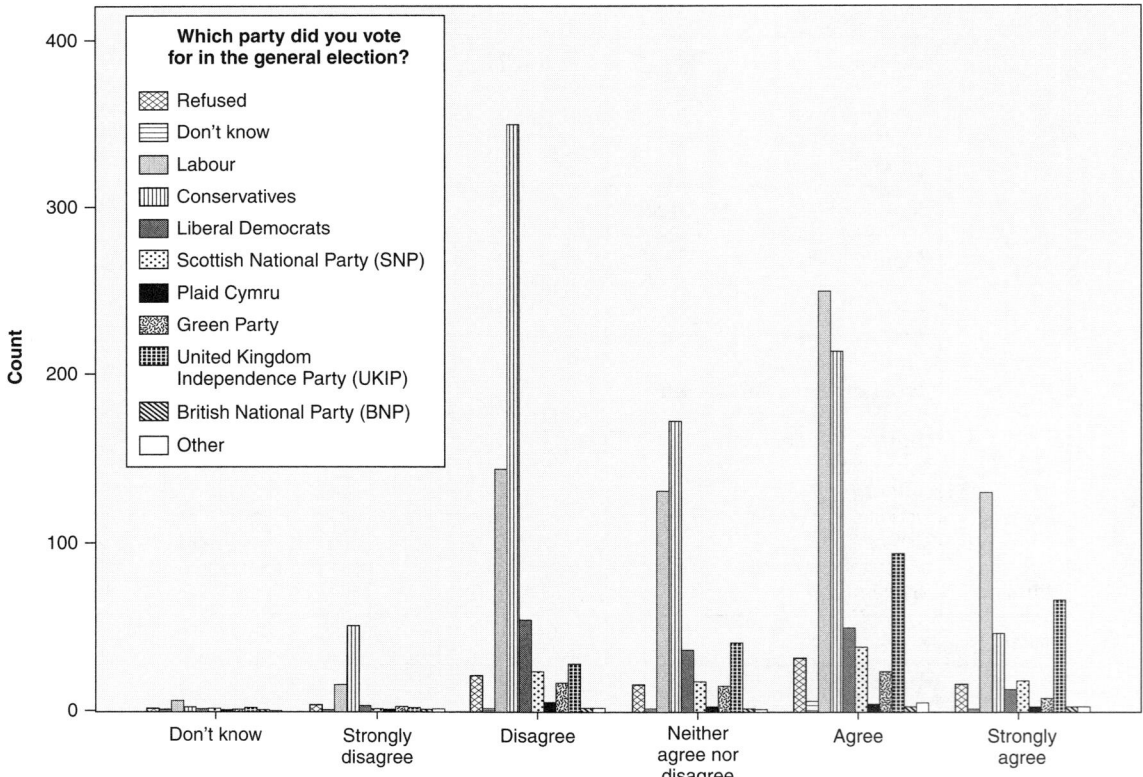

b02 *Which party did you vote for in the general election?*

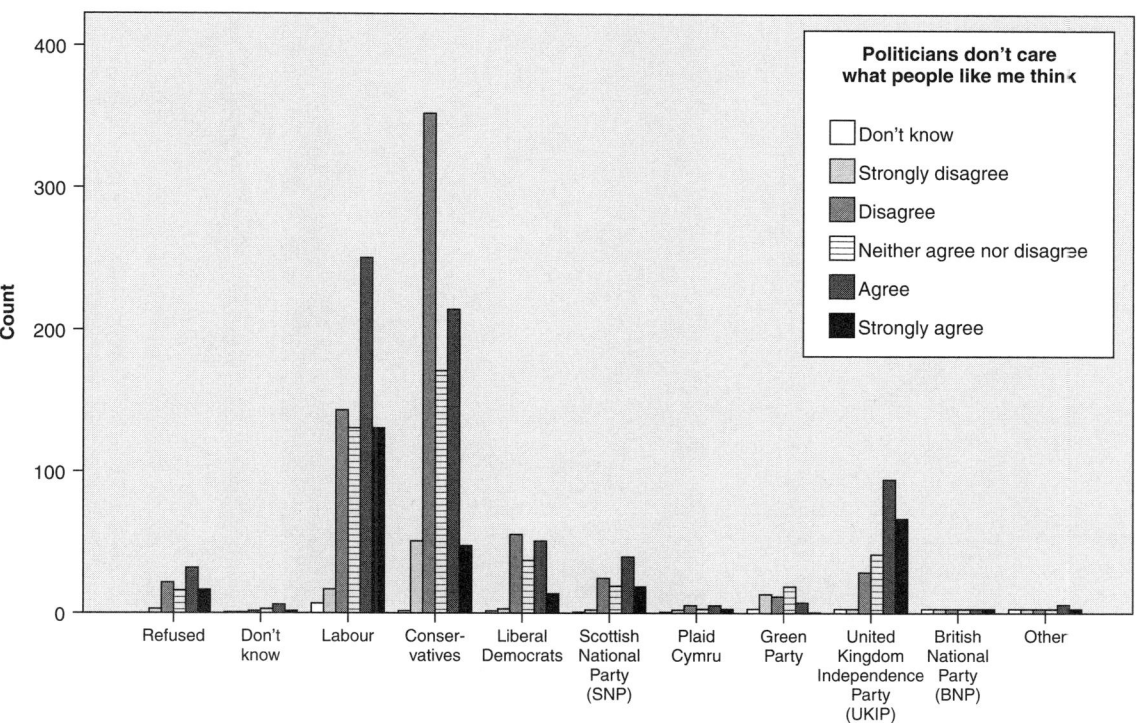

Which party did you vote for in the general...

JUST FOR FUN/CHALLENGE YOURSELF

1. When the cluster bar is produced with m02_1 *Politicians don't care about what people like me think* on the x-axis, the distribution is based on the political party the respondents voted for. The focus is on the parties, so if a researcher wanted to know specific counts by party in order to hypothesize about them, then the first arrangement would be best. When the cluster bar is produced with b02 *Which party did you vote for in the general election* on the x-axis, the distribution is based on the respondents' thoughts of politicians not caring about what people like them think. The focus is on how the respondents ascribe to the idea of politicians who do not care about constituents' thoughts. If a researcher wanted to know specific counts by respondents' answers about politicians in order to hypothesize about them, then the second arrangement would be best.